Content Audits and **Inventories**
A Handbook

Paula Ladenburg Land

The Content
Wrangler
Content Strategy Series

Content Audits and Inventories
A Handbook

Credits

Series Producer and Editor:	Scott Abel
Series Editor:	Laura Creekmore
Series Indexer:	Cheryl Landes
Book Editor:	Marcia Riefer Johnston
Series Cover Designer:	Marc Posch
Author Photo:	Anita Nowacka
Publisher:	Richard Hamilton

Disclaimer

Trademarks

XML Press
Laguna Hills, California
http://xmlpress.net

First Edition
978-1-937434-38-0 (print)
978-1-937434-39-7 (ebook)

Table of Contents

Foreword

The dreaded content inventory and audit…

To many of us, they are unnerving – a bit like a visit to the dentist. We know regular check-ups are important, but fear the outcome. Maybe we will have to face painful treatment, return trips and high costs to put things right. Then, before we know it, the next check-up is due and the stressful cycle starts again.

But just as looking after your teeth and consulting a dentist pay off, so do the content inventory and audit. They provide a systematic means to a valuable end and are a vital part of a core content strategy. Done regularly, they lessen the stress, allowing us to catch problems while they are easier to fix.

The rapidly-evolving, multi-disciplinary world of digital communication urgently needs to learn about content auditing to sort itself out. As organizations of all shapes and sizes struggle to work out their publishing guidelines, a massive array of content is being let loose across multiple channels and devices, for widely different purposes and audiences. Keeping track of it, let alone diplomatically evaluating, governing and planning it, is a major challenge.

How can we be sure that we are setting the right standards for our content teams to comply with? It's easy to say that content should meet our organization's business strategy and our audiences' needs – but often the two are hard to synchronize. The role of the content strategist is to bridge the gap and lead the way ahead.

Just as a single, unified content strategy has many integrated layers, working out a content audit is a highly collaborative task. We need input from a range of stakeholders to shortlist what to check for in different contexts. An audit that addresses all aspects not only results in a better customer experience but commits everyone to maintaining high quality content.

Often people ask: which comes first, the content strategy or content audit? Usually they work hand in hand. Before you start a content inventory or audit, you must know where you want to head strategically. An initial content audit will verify or uncover issues to address in the strategy. Once the strategy is in place, you can design follow-up audits to monitor performance and influence further tactics.

A few years ago, many of us working in the developing discipline of content strategy were anxiously trying to learn from one another how best to go about content inventories and audits. What their full scope was. What worked well. What went wrong. The analytical skills we needed to master. And the lions and tigers we encountered on the way.

Then along came Paula Land – a calm, cool, highly experienced voice of reason – who methodically drew all the complex strands together, introduced a time-saving toolkit of techniques, and spoke us through them clearly.

Whether you are new to the field, highly experienced, or somewhere in between, you are lucky to have immediate access to Paula's wisdom through this excellent book. She explains the "why" as well as the "what" and "how" of both a content inventory and an audit – and gives meticulous, step-by-step guidance to keep you, your organization or clients on a practical analytical track for continuous improvement throughout your content projects.

This is a handbook that removes the dread and leads you confidently ahead, with obvious returns on all your efforts. Keep it close beside you. I'm sure you will continue to refer to it for many years to come.

Diana Railton, DRCC[1]
Bath, UK

---

[1] http://www.drcc.co.uk/diana-railton

Preface

> If people are to be expected to put up with turning on a computer to read a screen, they must be rewarded with deep and extremely up-to-date information that they can explore at will.
>
> —Bill Gates, "Content is King"

Bill Gates is neither a content strategist nor a content marketer, but this statement from his 1996 blog post evokes definitions of what we now call content strategy: the analysis and planning related to delivering the right content to the right audience at the right time. "Deep." "Up-to-date." Those terms imply evaluation against some set of standards to assess depth, currency, relevance, and quality. To determine what is good, we need to know what bad looks like.

Advocating for Quality Content

As web content professionals, we have the opportunity and obligation to advocate for quality content. The tactics and strategies we use to create, publish, and govern quality content include the content inventory and audit, which, together, form the subject of this book.

A comprehensive content strategy is built on a foundation of thorough understanding and analysis of existing content, assessed against business goals, user goals, standards, and best practices. The first step in developing that analysis is the content inventory, a dive into existing content to understand the quantity, type, and structure. The second step, the content audit, builds on the inventory. When you audit content, you evaluate it against goals and standards, and you analyze it for quality and effectiveness, revealing information that can be used to improve existing content and plan for the future state.

Together, the content inventory and audit combine technology's ability to quickly gather and process data with the human brain's ability to use that data to analyze and strategize.

It's a Big Web Out There

According to a web-server survey[1] published by Netcraft, in March of 2012 (the last date for which information was available as this book was going to press), live websites numbered nearly 650 million. By the time you read this book, of course, that number will have increased. That's a lot of content. Somewhere out there, 650 million or so people are looking

---

[1] http://news.netcraft.com/archives/2012/03/05/march-2012-web-server-survey.html

at their sites wondering how to get a handle on their content – how to know what they have, how to know whether it's any good, and how to know what *good* even means.

Do you ask yourself those questions? If so, I have a proposal. Start by analyzing what you have. Start with a content inventory and an audit.

This Book's Audience

This book is intended for anyone doing any kind of digital project that involves an existing body of content. If you are a student of content strategy or a new-to-somewhat-experienced content strategist, site manager, or content owner charged with improving a website, this book is for you. If you're an experienced content strategist, you may find tips that you can add to your auditing toolbox. If you own a business website, this book may convince you of the value of monitoring your content and incorporating content strategy and governance into your organizational processes.

This Book's Purpose

This handbook introduces the concepts of inventory and audit in a business context, giving practical tips for putting data and analysis together to form insights that can drive a content project forward.

In my experience as a content strategy consultant, the inventory and audit are the necessary first steps to planning and implementing a content improvement project. Because these tasks can seem overwhelming, and because time and resources are always limited, it is important to take the time to plan, scope, and focus efforts for maximum value and reduce wasted effort. The strategies and tactics discussed in this book can help you make the most of your time and get the most valuable insights possible from your efforts.

As the web continues to grow, issues of content governance become even more critical. If we are going to develop and maintain high-quality websites, we need to pay attention to the full content lifecycle – not just development, but ongoing improvement and weeding. That's where this book comes in. The inventory and audit are valuable tools we can use to ensure that our websites are current, accurate, and effective. Adding the inventory and audit to your toolkit enables you to develop effective content strategies.

A Note About the Content Analysis Tool

In addition to working as a content strategy consultant, I am the cofounder of a software company called Content Insight. Content Insight developed and released a tool called the Content Analysis Tool (CAT), which automates the creation of content inventories. I developed this tool because I saw, over and over, my clients' need to alleviate the time-consuming drudgery of compiling their site data and their need to speed up their analysis.

While I discuss the Content Analysis Tool in Chapter 4, this book is not intended to serve as an advertisement or user guide for the tool. The information that I suggest gathering as part of a content inventory and audit can be gathered in other ways, either manually or by using other tools that provide some of the same functions and data sets.

For more on the Content Analysis Tool, see www.content-insight.com.

Acknowledgments

Without the community of content professionals who have defined and promoted the discipline of content strategy, this book would not exist. I owe a debt of gratitude to all who have shared their experience and wisdom and from whom I've learned so much.

Big thanks are also due the people who have provided thoughtful input and feedback on the book and encouraged its progress: Misty Weaver, Beth Bader, Kevin Nichols, Marcia Riefer Johnston, the late Emma Hamer, and of course, publisher Richard Hamilton and series editor Scott Abel. Thanks, too, to Diana Railton for graciously agreeing to write the foreword.

Special thanks to my husband Steve for his loving support and endless patience.

Introduction

> The more you seek to understand your content, the better your other work will be.
> —Sara Wachter-Boettcher, "Content Knowledge is Power"[1]

Writers want their ideas communicated. Businesses want their customers served. And readers want to be informed, entertained, or supported. These are the goals of an informed *content strategy*. The *content inventory* and *content audit* help achieve all those goals.

Content inventories and audits are methods of analyzing a set of content from both a quantitative (inventory) and qualitative (audit) perspective. Inventories and audits are usually done either as part of a larger content analysis and improvement project or as ongoing maintenance.

The inventory establishes scope and begins to reveal patterns in content quantity and type; the audit helps clarify and refine that scope, revealing a fuller picture of what needs to be addressed. Inventories and audits are a means to an end based on the principles that you can't improve what you can't quantify, you can't fix problems without identifying their root cause, and simply gathering data without analyzing it is an exercise in futility.

In this book, we begin by looking at how to make the case for doing inventories and audits, since many organizations lack experience with these activities and since getting permission and buy-in to spend time and resources may require some justification. Then, we'll look at how to put together an audit project, including assembling the team and establishing the goals and scope of the effort. After laying that groundwork, the following chapters dive into the various methodologies for auditing content, from qualitative to competitive. We finish up with some advice on presenting findings to stakeholders.

What exactly do I mean by these two key terms, content inventory and content audit? Here are my definitions.

The Content Inventory

A content inventory is a quantitative assessment of all the content on a website – a list of all the pages, images, and other files that make up the content set as well as data associated with those files, such as content type and metadata.

---

[1] http://www.smashingmagazine.com/2013/04/29/content-knowledge-is-power/

The inventory provides summary information about the site, which is helpful for establishing scope. It also provides detailed information about each page on the site, including analytics data, lists of links in and out from each page, the metadata (title, description, keywords), H1 text, word count, a list of images on the page, lists of audio or video on the page, and any documents, such as PDFs. Although in common use it usually refers to the content of a website, a complete inventory of all an organization's content assets may also include other digital content such as newsletters or email campaigns and print content, such as catalogs.

In the course of a redesign or site migration project, the content strategist or content manager will supplement the basic data about the content with other information relevant to the project. This additional data may include content ownership, review status, content and template type information, notes for migration, redirects, and SEO-optimized URLs. The inventory is often organized by the structure of the site so that a site navigational model can be derived. The inventory may also be used as the method of tracking content from one system to another.

A content inventory gives you a foundation for moving to your content analysis project's next step: the content audit.

Inventories can be created manually or with a tool or combination of tools. Inventory creation is discussed in Chapter 4.

The Content Audit

A content audit is a qualitative evaluation of a set of content. When you audit content, you assess it against a variety of measures depending on your context and goals. Typical audit tactics and criteria, as discussed in this book, range from editorial issues (the content's consistency with brand and messaging guidelines, for example, and its quality, depth, and breadth) to audience appropriateness to performance against competitors' content to effectiveness as measured by analytics data.

By transforming a content inventory into a content audit, you gain a powerful tool in further understanding your content's strengths and weaknesses. Containing both quantitative and qualitative data, a content audit allows you to dig deeper, analyzing page-by-page how your content is structured, displayed, and maintained, as well as what value it provides to your business and your users.

When Do You Inventory and Audit?

Moving a website from one platform to another, redesigning a website, creating a unified content strategy,[2] or making any other major change to a website is like moving into a new house or office: you have the opportunity to take stock and remove waste, labeling and evaluating the usefulness of everything you take with you to the new location. Once you have the content in place, you can avoid accumulating outdated content by doing a rolling content audit, a regular review of content to keep it fresh and make the next redesign or move easier.

Following are some of the specific reasons that trigger an inventory and audit and the ways in which the focus may differ.

Site Redesign

Usually you perform a content inventory and audit during a larger site redesign. In this situation, if the site is moving to a new platform, you need to complete the entire process early enough to ensure that the findings are incorporated into the information architecture, the page templates, the creative design, and the configuration of the *content management system* (CMS).

The inventory provides the information necessary to scope the project and set the baseline against which the future state will be measured. The audit should address issues of content currency and quality as well as inform a gap analysis, whether you are assessing the content against the organization's goals, against competitor sites, or both.

Before a Migration

Most sites at some point are migrated from legacy systems to newer platforms to take advantage of new technology and improve content management.

When you know the current size and state of your website, you're in a better position to evaluate new platforms or CMSes. If you plan to outsource the migration and implementation work, accurate knowledge of your content helps simplify the scoping and RFP (request for proposals) process and aids in evaluating systems against your content needs.

---

[2] http://www.managingenterprisecontent.com

Updated inventories are useful from several perspectives:

- **Stakeholders and project managers:** defining scope, timelines, and budgets
- **Technologists:** determining system needs and requirements
- **Content owners and managers:** defining content needs and workflows

Regularly updating your content audit helps you keep your site migration-ready all the time. The updated audit reflects site growth and change and provides an at-a-glance resource for planning and managing a migration project.

Ongoing

Noted content strategist and author Ann Rockley, in her keynote at the Confab 2012 conference, compared maintaining a website to tending a garden. To thrive, every site needs regular watering and weeding. Governance practices that enable you to identify areas needing attention help create an environment in which your content can flourish. When you track content changes, you have a picture of how fresh your content is and how often it's updated so you know when to "water and weed."

In other words, ideally an audit is not a one-time project but a repeatable and regularly scheduled process. Whether you're migrating a site or just maintaining it, taking the time to audit regularly will help you provide the most consistent, current, and relevant content to your users.

If you keep track of content changes, you can keep your content inventory up to date as part of regular website maintenance and management. An automated inventory can make this process easier by providing a change report (see Chapter 4, for more on tool-supported inventorying). By making it easier to see the *content lifecycle*, a regularly updated inventory gives your team a powerful tool for communicating website changes.

How Frequently Should You Audit?

Ultimately, the answer to the question of how frequently you should audit is "as often as possible." As mentioned above, a rolling audit saves time in the long run, since content is actively managed and the opportunity for *ROT* (redundant, outdated, or trivial content) to build up is significantly reduced. As the saying goes, "An ounce of prevention is worth a pound of cure."

Several factors – including *governance*, publication frequency, currency, and accuracy – influence the question of how frequently you should audit your content.

Governance

If your organization has robust governance policies in place that ensure accuracy and quality of content as it is published, you may need to audit less frequently. But since each piece of new content is published into an existing content set, you still need to audit regularly unless someone always reviews the related content at the time of publication to check for consistency, lack of duplication, good cross-linking, and overall fit within the content set.

Publication Frequency

The frequency with which content is published also affects how often you should audit. If the site is rarely updated and most content is static and "evergreen," that is, if the content is unlikely to fall out of compliance with brand, legal, regulatory or other standards and guidelines – and if it's likely to remain relevant to customers – an annual or semiannual audit may be enough. But few websites meet that description. Most are updated frequently, and each update introduces the possibility of duplication, error, or other issues. Audit as frequently as resources allow. If necessary, scope the audit to focus on the newest content and the content surrounding it or most relevant to it.

Currency and Accuracy

Content that's ephemeral – for example, seasonal, event-related, or price-related information – should be regularly weeded. For this type of content, you may want to use your CMSes automation capabilities to remove expired content or follow a governance policy that requires review and, as appropriate, removal on a schedule.

What Content Should You Audit?

A thorough content inventory and audit spans all of an organization's content, whether digital or print, on all channels. This book focuses on website content, but many of the strategies for assessing content quality could be adapted to non-web assets. A successful content strategy ensures a consistent level of quality and consistent messaging across all user touchpoints, whether via a website or other digital communications, such as newsletters, email campaigns, print materials, social media, and syndicated content. You can apply the guidance in the book to any type of content.

Why Quality Matters

Imagine that you're opening a restaurant. Your architect has designed an impressive space. Your interior decorator has chosen a beautiful color palette and attractive furniture and décor. But your architect and designer know nothing about your customers. They haven't thought through the way people move through a space or what makes them feel comfortable. The entrance to the restaurant is unmarked and inspires no confidence in what's on the other side of the door.

Worse yet, the kitchen staff aren't prepared. They have recipes, but, with no idea who the patrons will be, they don't know what to make. Are they cooking for connoisseurs of fine cuisine or seekers of comfort food? They don't know. They don't even know what ingredients and equipment they have to cook with. How fresh is that produce? Is there enough flour to make the bread? Will that oven get hot enough for pizza?

Your chefs can't answer those critical questions.

The front-of-house staff are equally clueless. The waitstaff don't know what the dishes are or how to describe or present them. They don't know what order dishes should be served in or what wine to serve with them. Baffled patrons are seated haphazardly throughout the restaurant, so one server has to take care of the majority of the tables while the others stand idle.

Would you eat at a restaurant this ill-prepared? Would you ask someone to spend $50 or $100 on a meal here?

Now imagine that we're talking about a website's user experience and content.

As user experience and content professionals, we are in the business of designing sites that are well-structured, useful, and usable. A customer who spends money on a meal that fails to deliver is disappointed. But when we serve up websites that fail to deliver – and then present a client with a bill for six or seven figures – the consequences are much more serious, including the risk of lost business and even legal action.

How do we avoid that disastrous situation? By spending the time up front to do the necessary research, ideation, and planning. We work closely with all the elements that go into delivering a high-quality content experience, and we set goals for and measure our results. The content inventory and audit are the tools we use to do that research and strategy development so that what we deliver serves our customers while supporting our business goals.

Laying the Groundwork

It is tempting to just jump in and start an inventory and audit, particularly if you have a tight project deadline. Before you begin, taking the time to set the stage for a smooth-running, comprehensive project will help ensure that the data and insights gathered will result in useful, usable outcomes.

In this section, learn how to define your business context, plan for project management, select your team members, choose your tools, and scope your effort. Learn the basics of the content inventory and how to build it out into a comprehensive content audit.

CHAPTER 1
Building the Business Case

A business asset is something that is of value to a company. Your web content can be an asset, but you'll need to explain how and why before you can expect to gain stakeholder buy-in and support.

—Kristina Halvorson, *Content Strategy for the Web*

In This Chapter

Before you begin a forward-looking website content project, you often need to justify the preliminary work of measuring and analyzing the current state. Making the case for investing time and resources in a content inventory and audit involves showing the value of these activities by putting them in a business context.

A Framework for Improvement

There is an analogous framework for the inventory-audit-analysis process in the business-process improvement methodologies that are part of Six Sigma. The *DMAIC framework*[1] (define, measure, analyze, improve, control) is used for improving, optimizing, and stabilizing business processes and designs.

Here's how you might adapt the DMAIC framework to a content project.

Define: Document customer needs, business goals, and project goals. Review with stakeholders and get organizational buy-in before beginning.

Measure: Collect the baseline set of data. Determine which aspects of current processes and current metrics for content performance and user engagement you will analyze, and collect the data for the initial baseline. At the end of the project, measure updated data against the previous data to show improvement.

Analyze: Look for cause-and-effect relationships, find root causes, list and prioritize problems and potential solutions. Map process flows. Plan content improvements.

Improve: Building on what you learned during the analysis, optimize current content and processes. Pilot new processes and workflows.

---

[1] http://www.isixsigma.com/dictionary/dmaic/

Control: Future-proof the content by establishing standards and guidelines, training personnel on new processes, and putting measurement plans into place. Plan for continuous monitoring and optimization.

This book focuses primarily on the Define, Measure, and Analyze steps. Improving content and processes are the follow-on activities enabled by an inventory and audit project, and control – the ongoing maintenance and governance of content and content processes – is best built on a foundation of deep knowledge of content strategy goals and performance metrics.

"Define" = Establish Your Content Goals

In the DMAIC framework, the *d* stands for *define*. In the context of content inventories and audits, the Define step equates to establishing your content goals.

Why Are You Auditing?

Content projects exist in a business context. Whatever your reason for auditing, the larger reason is that the business wants the content to achieve something. Does the content fulfill core business and user goals? The audit answers that question. You also want to answer these questions: What do the missed opportunities cost? What does addressing these issues cost? How do we fix the issues we've identified?

When you quantify those goals and reduce them to a statement of intent, you have a touchstone to refer to after you've gone through the content details. Your goal statement is like a mission statement – a higher-level purpose that informs the strategy.

For example, your goal statement might be something like this:

> To remove redundant content, create a consistent voice, present clear calls to action, and simplify the purchase funnel.

You may even tie your goals to metrics:

> Migrate only the content that consistently performs in the top 50% of our traffic data and reduce the registration process to three steps.

What Do You Need to Learn?

Depending on your role and the project, you may focus your audit in different ways, including the following possibilities.

Identifying the Site's Structure

Are you creating an inventory to identify the structure of the site and the content? Perhaps you're an information architect who needs a sense of the structure – what is the navigation model, how is content grouped and classified, how many *interaction models* or *templates* does the team need? Can any content be reached only via a text link buried deep in a fourth-level page? If so, you would probably begin by organizing the file list into a *site map*, reflecting navigational models and site hierarchy, and looking for patterns and discrepancies.

Seeing the Big Picture of the Site Content

Are you a content strategist preparing to assess the content itself? In addition to the structure of the site, you may be assessing the content types and the relative quantity and depth of types – is the site heavy on marketing copy but light on informational copy or help?

If you're migrating a CMS, you want to know where the content lives, who owns or manages it, and how it gets published. To help decide which content to migrate, you would evaluate content against your organization's important criteria. At the least, you evaluate currency, messaging, clarity, consistency, and user-task facilitation – in short, the qualitative aspects of the content.

Managing Content Quality and Performance

Are you a site manager trying to keep your site performing well and free of orphaned pages, outdated content, duplicate content, and broken links and forms? Are you tracking to a *content migration* and want to make sure you've accounted for everything you need – and don't need – to carry over to the new system? Tracking the differences in your site from day to day or week to week is critical to making sure that everything is running smoothly. You're probably also interested in mapping your inventory against site analytics and *SEO* guidelines.

Whatever your focus, any audit should include these basics:

- **Content disposition:** what to keep, what to remove, what to revise, what to create?
- **Content ownership:** who is responsible for making decisions about each piece of content?
- **Content quality:** how well does each piece perform its intended function?

Understanding Stakeholder Goals

Conducting *stakeholder interviews* before beginning your audit is a valuable exercise. The stakeholders are the people you are working for, and their insights and pain points set the context for the audit process.

Talk to the executive sponsors of the project, the business owners of the content, the people who create and publish it, and the people who maintain and support the CMS and the site. In short, talk to anyone who has a stake in the site's success. Stakeholder interviews, besides setting the overall context for the audit, are also great inputs into *persona* creation. If you can interview site users and internal staff, you can create composite personas against which to measure your insights and recommendations.

When you talk with the stakeholders, ask whether they have already attempted content initiatives and, if so, how they went. Knowing what the team has been through before – what worked and what didn't – can help you avoid going down the wrong path and can show you how to work with the stakeholders. If they're in a "once bitten, twice shy" frame of mind going into the project, you may have to do more organizational-readiness work and manage your communications accordingly.

Many templates are available on the web; the basics are the same. Your template should address, at minimum, these issues:

- Roles and responsibilities
- Goals and objectives for the site
- Users
- Site-publishing workflow specifics – who, what, when, how often
- Pain points – internally and from the customers' perspectives
- Success metrics

Related: Appendix B, *Stakeholder Interview Template*

"Measure" = Inventory Your Content

In the DMAIC framework, the *m* stands for *measure*. Inventories help us measure the as-is state of a website and provide the basis for later steps in the process, such as scoping a project or doing a content audit.

Think about the reasons retailers take inventory. They need to know what is selling and what isn't, what to reorder, what they have too much of, and whether something isn't selling because it's in the wrong location. A content inventory provides that same detailed look into the content assets you have to work with as you begin your content project.

Why spend the time to gather this data? Wouldn't it be easier and more fun to just jump right in and start looking at the content?

We create content inventories for lots of reasons. The core purpose of an inventory is to give you a starting place. In most content projects, you're trying to determine how to get from the current state to a desired state. The inventory enables you to define your starting point.

Inventories also serve another purpose in helping sell the proposed project to an organization. Few people in an organization have a handle on how much content is on the website and what it is. Each department or content stakeholder may have an idea of their section but may not know what else is out there. Content often exists in organizational silos, and the content auditor – the person conducting the inventory and audit – may be the only one looking across all those silos.

When you get to the audit portion of the project, where qualitative judgments are made, it's possible to ruffle some feathers. People often defend their silos. So a benefit of a content inventory is that it depersonalizes numbers. You can defuse some potential conflicts by simply gathering people around the idea of moving forward in ways suggested by the data. A dispassionate presentation may inform the stakeholders that clearly addressable issues are supported by data.

For example, stakeholders may learn that the site has more pages than they realized; that none of them have current dates, metadata tags, or customized page titles; and that some pages of the site receive no traffic. As Joe Gollner of Gnostyx Research has been known to say, "Good data makes for good conversations."

Content inventories can be a hard sell. Presenting a stakeholder with a spreadsheet full of painstakingly-gathered data about the website may not seem like a compelling way to make a case for a qualitative project. You may want to pull out some of the most interesting or unexpected results into a graphic presentation.

In a recent project, for example, my clients were surprised to see that the bulk of their site's content was directed to an audience that they did not consider primary. The primary audience, on the other hand, was comparatively underserved. Quantity does not equal quality, of course, but a statistic like that can be an indicator that organizational resources are not well aligned with business goals.

Related: Chapter 14, *Presenting Audit Findings*

"Analyze" = Audit Your Content

In the DMAIC framework, the *a* stands for *analyze*. When you audit a website, you analyze it page by page against a set of criteria. The audit can take weeks or even months depending on the size of the site and the goals of the audit. A later chapter discusses how to limit the scope of an audit if you don't have time to assess each page in depth. First, let's look at the value that an audit provides and why you should argue for enough time to complete one when you're planning a project schedule.

Occasionally, someone associated with a website content project asks, "Why do we need to do a content inventory or audit?" Perhaps it's an account manager saying, "That sounds like discovery, and clients don't want to pay for discovery. Can't we just figure it out when we get there?" Or maybe the client has been sold a creative design and your project manager is saying, "We just have to make the content fit – the client loves the creative." Or maybe one of your colleagues or the client is arguing that "We're creating something brand new – we don't want our creativity constrained by what was done in the past."

Unless the plan is to completely replace all content on the site, every staff member involved with creating and managing it, and all the technology used to build and publish the site, you will be working within the constraints that existed during creation of the current site. Understanding how that current state was arrived at helps you detect areas for improvement – and identify what *is* working – which informs your strategy.

Content audits are critical to understanding and evaluating the performance of your content against business goals, editorial standards, user needs, and performance factors such as search engine optimization and content use or web analytics. They bring value to your website project and ongoing maintenance tasks by enabling you to catalog and analyze your content structures, patterns, and consistency.

An audit tells you what needs to be done to bring the existing content into conformity with brand, editorial, and other standards your organization or client has set, and it ensures that the content fulfills business goals and meets the customers' or users' needs. Tailoring a content audit to your organization's content goals enables you to focus on the factors that return the most benefit.

Content audits enable you to do the following:

- Identify whether content consistently follows template, editorial, style, and metadata guidelines
- Assess whether the content supports business and user goals
- Establish a basis for analyzing the gap between the content you have and the content you need
- Prepare content for revision, removal, and migration

"Improve" = Making the Case for Change

In the DMAIC framework, *i* stands for *improve*. The inventory establishes scope and begins to reveal patterns in content quantity and type; the audit helps clarify and refine that scope, revealing a fuller picture of what needs to be addressed. Inventories and audits can serve as communication tools throughout a project's or website's lifecycle. These activities and their outcomes can provide a way to connect stakeholders, designers, content managers, and technologists and gather the team around the implications and opportunities for change.

A key point here is the importance of understanding the organization's goals and the role content plays in achieving them. Without that background, it is difficult to focus an audit for maximum impact, and the results will be less compelling to the audience. *With* that background, you can more easily find content improvement opportunities that people in the organization will support.

By the end of a content audit, you will probably know more about the content on the site than anyone else in the organization. You may be the only person whose knowledge spans all the organizational content silos. This puts you in a position of power. If you can back up your story with data and informed analysis, you can move change forward.

"Control" = Return on Investment

In the DMAIC framework, *c* stands for *control*. Creating a systematic way to measure improvement and govern ongoing processes completes the picture. Defining and quantifying the *return on investment* for making the recommended changes helps gain organizational buy-in.

Proving the need for change to the stakeholders may require more than revealing the content issues you've discovered. You may need to provide justification based on numbers that show the hard costs of missed opportunities and process inefficiencies. Can you calculate how much your time is worth? How much is your customer's attention and loyalty worth?

What is the competitive disadvantage incurred by having outdated, inaccurate, or unengaging content?

Few of us would willingly sign up for creating or implementing a strategy that isn't informed by the existing state of the content. Knowing what assets we have to work with, understanding the environment in which they were created, and knowing the goals they were intended to support grounds us in the business context. Understanding what is and what is not working, of course, helps us address issues most effectively.

Your organization's decision makers need to understand that a comprehensive, unbiased assessment of the current state of your site's content is critical to the creation of an effective content strategy and justifies the costs of doing the time-intensive inventory and audit. Business justification usually relies on making a case that the proposed initiative will help the company save money, make money, or both. For example, you might quantify potential benefits like these:

- Increased sales
- Faster time to market due to easier product content deployment
- Process efficiencies, resulting in reduced personnel costs
- Greater customer satisfaction
- Reduced legal risk

If you need to convince the decision makers to allow you to spend time and resources on a content inventory and audit, think through how your organization generates revenue or otherwise shows value. If you can prove that the findings and outcomes of an inventory and audit will help the organization's bottom line, it will be easier to get approval.

The costs and benefits can be both "hard" – such as increased sales, process efficiencies, reduced production costs, and fewer calls to customer support – and "soft" – such as greater customer engagement and brand consistency. When you have established the goal, doing some tests of the current experience and projections of what they may be costing the organization could help you make the case.

For example, in an online retail site, you might demonstrate how the quality of the product content and the user experience of making a purchase affects customer purchase behavior. You might set up a simple, low-cost usability study of the current site with potential customers (invite friends and family if necessary) to show where the gaps and poor experiences are hindering sales. On a nonprofit site that solicits online donations, you might show that an ineffective call to action or a convoluted donation transaction path discourages potential donors.

By calculating the time it takes to regularly inventory and audit your content, as compared to the time required to do a major overhaul later – or the cost of losing customers and sales – you should be able to justify the expenditure of resources. Using automated tools to regularly assess and track your content is a quick way to show cost savings over manual efforts. Finding issues early and addressing them as they arise prevents content from getting out of control and failing to deliver on business and user needs.

Summary

Content inventories and audits provide a valuable window into the current state of your organization's content. When you inventory and audit within a framework of business goals and measurable outcomes, you help drive positive change. The time it takes to think through and document your goals and to get stakeholder buy-in at the beginning of your project is time well spent; it ensures that you've captured the right kinds of data and have a clear sense of how you will use the information you glean.

Why Inventory?

- Assess details of a site or content set
- Scope a project for resource estimation
- Identify patterns in content structure
- Set a baseline to measure future site against (ROI)
- Establish a basis for tracking content through a migration

Why Audit?

- Assess the current state of content to inform strategy
- Identify whether content consistently follows brand, template, editorial, style, and metadata guidelines
- Assess whether content supports business and user goals
- Establish a basis for gap analysis between content you have and content you need
- Prepare content for revision, removal, and migration
- Uncover patterns in content to support structured content initiatives
- Develop a deep understanding of the content

CHAPTER 2
Planning an Inventory and Audit Project

Start with the end in mind.
       —Stephen R. Covey, *Seven Habits of Highly Effective People*[1]

In This Chapter

You can improve your chances of getting buy-in for an inventory and audit project by showing up front that you have a carefully thought-out plan and a solid project management approach, and by regularly communicating your progress and findings. This chapter presents a brief overview of how to plan and manage your project.

Putting Together Your Project Plan

A content inventory and audit project can be a time- and resource-consuming effort. Taking a project management approach to your inventory and audit activities can provide a helpful structure within which to work and ensure that all the necessary input has been sought and activities completed. The rigor of an organized project process can also help you demonstrate value to your stakeholders and ensure that your outcomes are as actionable as possible. Begin by explicitly stating the goals for the project, select and prepare the team members who will be involved, and plan for how you will track and communicate progress.

Document Your Vision and Goals

Explain the vision and goals of the project to your stakeholders and to your project team to make sure that everyone is working toward the same end. Ensure that the team agrees about who the audience for the audit is, what business goals it supports, and what the project requirements are. Establishing these goals helps everyone on the team make cohesive decisions and spend time effectively.

Define Roles and Responsibilities

Inventories and audits are ideally the work of a cross-disciplinary team that assesses the content from a variety of angles. Analyze your team to see whether you have sufficient content strategy, user experience, business and marketing, and technical representation. Note that these roles may not always be filled by different people; in a small organization, people

---

[1] https://www.goodreads.com/work/quotes/6277

typically wear several hats. The point is that you want to have as many relevant viewpoints on the process as possible.

Identify who will do what on the project to safeguard against duplicated effort or important tasks falling through the cracks. Make sure that everyone on the team is aware of his or her role; understands the dependencies on other team members; and is clear on the timing, format, and content of expected deliverables. You may want to create a swimlane diagram or a *RACI* grid (which identifies who is *responsible, accountable, consulted,* and *informed*) to document roles. And you may want to do a walkthrough of the project lifecycle showing what happens when.

Related: Chapter 3, *Assembling the Team*

Choose Your Tools

Inventories and audits can be done manually or can be supported by tools. Whether or not you use a tool to generate your inventory, you will likely be working with software to create your project plan, write your documents, and visualize and present your findings. You may also need some familiarity with your current or future CMS. Before you start, select the tools you will use for each of these steps, and confirm the following:

- The tool scales to your project size and requirements.
- The tool does everything you need it to do.
- Everyone who needs to use the tool has been trained and is comfortable using it.
- The tool meets your organization's security requirements (that is, your IT department has approved any new installations).
- Any appropriate fees are paid, and licenses have been attained.
- Workflows have been defined, documented, and communicated.

Related: Chapter 4, *Creating a Content Inventory*

Communicate Early and Often

Regular communication of project status, both to internal team members and to project stakeholders, helps everyone feel comfortable that things are on track. Regular check-ins also create a place for issues and risks to be raised and addressed early. For example, during the course of the inventory and audit, you may discover unexpected information – hidden pockets of content or technical issues that would affect scope and timing. Better to know and plan for these scenarios early and often.

Related: Chapter 14, *Presenting Audit Findings*

Set Milestones and Exit Criteria

When do you stop? When you are auditing content, it can be difficult to know when you have collected enough information to inform your actions. This is why you need to understand your business goals, set milestones for communication of results along the way, and define your audit scope. Stop when you have enough data and analysis to support recommendations that you feel can result in measurable improvement. Document and present your findings.

Related: Chapter 14, *Presenting Audit Findings*

Summary

Approach your content inventory and audit activities as a defined project with clear goals, defined roles and responsibilities, the right tools, and a mechanism for regularly communicating progress and findings. A well-defined project is also a tool for convincing stakeholders of the value of investing the resources.

CHAPTER 3
Assembling the Team

Developing a content strategy that helps you lead a team or project through the process is, from the perspective of a decision maker, a way to maximize your ROI on content investment.

—Rahel Anne Bailie and Noz Urbina, *Content Strategy: Connecting the dots between business, brand, and benefits*

In This Chapter

The content on a website is part of an overall *user experience* (UX) and business context that has many owners within the organization. Involving some or all of them in the inventory and audit process helps ensure that different perspectives are sought and that all are working toward shared goals.

Who Should Audit?

Who within your organization or your client's organization should be responsible for auditing your site content? Anyone with a stake in how content is created, organized, and displayed can conduct a content audit, and the findings should be shared widely throughout the organization.

An individual can conduct a content audit. The process can be improved by formation of a multidisciplinary team. The goal of a content audit is not to simply collect data but to have the information you need to make good decisions. By breaking down the audit into areas of expertise and engaging a team who know the technology, systems, and standards involved in your website content lifecycle, everyone who is part of the process gains greater insight into the content.

Given that a content strategy initiative is often part of a larger project to improve the overall user experience and to upgrade technology, the cross-dependencies between these roles make a strong case for including members of all affected teams in the process, whether as participants in the audit, recipients of the information, or partners in decision making.

While the content team is conducting the inventory and auditing content against business and user goals, the UX team may be auditing site architecture and user flows, the SEO analysts may be running the numbers on content performance, and the tech team may be assessing the template and component requirements for the new content management system (CMS). These activities can run in parallel as long as all team members are in regular communication about progress and findings. An organized project, with a team member acting as project manager, can help facilitate managing against milestones and goals.

Related: Chapter 5, *Preparing for a Content Audit*

Content Strategists, Content Marketers, and Content Creators

Members of the content team – content strategists, content marketers, content creators – are the usual owners of the process and of the inventory and audit results. These team members do inventories and audits because those pieces support the strategy creation. Audits set the baseline for the current state of the content, help identify areas of low quality or poorly performing content, and inform a *gap analysis* and a roadmap of how to get from where the team is to where it wants to be.

Information Architects and User Experience Designers

Information architects and user experience (UX) designers also need the insights provided by an inventory and audit. Designing navigation systems, *user flows*, and *taxonomies* without the benefit of a thorough understanding of the existing content can lead to designs that don't adequately support user needs or behave appropriately in *multichannel* and *multidevice* environments.

Website redesigns are sometimes driven by the creative or UX teams. Sometimes this results in content being an afterthought or being brought into the project after templates have been created. This can result in a large amount of rework later when content doesn't fit the templates and either the templates have to be redeveloped or the content reconfigured. These headaches and expense can be avoided if all affected disciplines are on board and working together from the start.

Content Managers and Site Managers

For the site management and technical teams planning a content migration, it is critical to understand exactly what content exists, how much of it there is, and what might need to happen to it on the way from one system to another. A content audit done in advance of a website migration should include decisions about what content to keep (and revise, if needed) and what to remove, to ensure that only the content worth migrating is migrated. It should also address issues with the content that can be fixed as part of – or in advance of – the migration itself, such as coding issues that affect rendering or inconsistently structured content that will be harder to migrate cleanly. Audit findings often have an impact on the design and production of features within the CMS as well, such as the content input and rendering templates, the content tree structure, and the taxonomy and tagging features.

Sharing your audit results with the technical team – and pairing with your UX counterparts to ensure that the new CMS or site structure is appropriately configured for your content – saves headaches down the road.

SEO Analysts

The content inventory and audit are also powerful tools for monitoring site content and activity against SEO standards and analytics goals. An inventory that includes metadata associated with each page (titles, descriptions, keywords), H1 text, and analytics data gives the SEO analyst a quick way to identify what content to dive into for deeper analysis.

Customers

It may seem strange to say that customers should be site auditors, but the reality is that they already are. Every time someone interacts with your site, whether reading an article or buying a product, he or she forms an opinion. Watch your site analytics to see how long visitors stay, whether they return, and when and where they leave. These are all valuable clues to your content performance. Search logs tell you whether your visitors are speaking the same language you are and whether they're finding what they're looking for.

If you aren't paying attention to this data, you may be missing opportunities. And if visitors comment on your content, or if they review it or share it, you have direct evidence of their interaction and opinion. If your site doesn't include commenting, reviewing, or social sharing, you can always go directly to the source – set up user forums or one-on-one usability tests and expose customers to your content and get their feedback. Your customers are valuable auditors. Find as many ways to listen to them as possible and take their feedback seriously.

Roles and Responsibilities

What if you don't have a dedicated, multidisciplinary team to work on the audit? If your resources are limited, it may be necessary for team members to play several roles during the course of the project, changing focus as needed and taking on tasks that require skills typically found in other disciplines. In that situation, it becomes all the more necessary to have a clear set of guideposts and guardrails – well-defined tasks, clear scope and goals, and access to all the relevant guidelines, standards, and project requirements. It may be helpful to set up a responsibilities matrix using one of the standard methodologies such as RACI (who is *responsible* for completing the task, who is *accountable* for signing off on the work delivered, who needs to be *consulted* for subject matter expertise and input, who needs to be *informed* as to progress and outcomes).

A sample RACI matrix for an audit project might look like Table 3.1:

Table 3.1 – Sample RACI matrix

Activity	Content Strategist	Info. Architect	Analytics	Developer	Creative	Project Sponsor	Project Manager	Business Owner
Create inventory	R	A	C	C	I	I	A	I
Audit content	R	C	C	I	I	I	A	C
Template/ component audit	C	R	I	R	C	I	A	C
Brand audit	C	I	I	I	R	I	A	C
Design audit	C	C	I	I	R	I	A	I
Content re-commenda-tions	R	C	C	I	I	I	A	I

As valuable as the multidisciplinary team is during the content audit process, they can be just as valuable for the ongoing governance of content. The audit team is critical to making content decisions throughout the project, but those same perspectives and skills should be used going forward as well to ensure continuous oversight. Consider setting up regular meetings of the audit team to participate in a rolling audit to evaluate and optimize on an ongoing basis.

Related: Chapter 15, *The Ongoing Audit Process*

Summary

Involving everyone who has a stake in the outcome of an inventory and audit project is an effective way to distribute the workload and ensure full coverage. Involve members of the content team, the UX team, analysts, and technical people. Your customers are auditors too. Define roles, responsibilities, and dependencies, and regularly communicate on your progress and on issues that arise.

CHAPTER 4
Creating a Content Inventory

Not doing an inventory is like starting to bake when you don't know what
ingredients you have in the house.
—Rahel Bailie, Intentional Design blog[1]

In This Chapter

This chapter discusses what data about your content you might want to
include in an inventory, how to gather that data, and next steps.

About the Inventory

A content inventory is a detailed representation of the data about a
website or content set, often presented in the form of a spreadsheet. Inventories can be created manually, by navigating through a website,
visiting each page and copying the information into a spreadsheet, or
by running an automated tool and using the returned data as your
starting point for your eventual audit.

Whether you create your inventory manually, use a tool, or both, there
are essential data elements to capture for each page. The process of creating and organizing this data set will no doubt begin to spark thoughts
about the site structure and content patterns and will help identify the
issues to be delved into more deeply in the audit process.

At minimum, a content inventory contains a list of all the files on the
site – HTML pages, images, documents, media – and data about those
files that is helpful in the analysis stage. An inventory that includes data
such as the page meta title, description, and keyword enables the site
auditor to quickly assess the value of that metadata for supporting search
and content management. Layering in site analytics, such as page views,
provides the opportunity to find high- and low-performing content.
Links to and from each page indicate cross-linking strategy and provide
a list of links to update if the site structure changes. Word count aids in
scoping localization and enabling comparisons across similar content
types to see whether the depth is consistent. Appendix A contains an
example of a typical content inventory.

---

[1] http://intentionaldesign.ca/2012/08/09/content-inventories-audits-and-analyses-all-
part-of-benchmarking/

Basic Elements of an Inventory

While the data gathered in an inventory will vary depending on your content set, your projects needs, and your goals, most inventories have certain basic elements in common, as detailed below.

URL

Looking at a resource's URL structure lets you evaluate several things.

Length and clarity: Shorter URLs are better for both human readability and search engine optimization (SEO). A long URL may not be rendered by some browsers and is not memorable to a human who may later want to directly type it in. It's also a best practice to use hyphens (rather than underscores or blank spaces) between words in a URL; a quick look at the URL list helps you identify whether your URLs follow this practice.

Identifying and addressing poorly constructed URLs is not only a favor to your human users but gives you the opportunity to improve your site's ranking. URLs composed of session IDs or other parameters provide no information to the user to help set expectations of the content he or she is likely to find at that location. Multiple parameters may also affect whether a page is crawled by search engines like Google.

Navigational structure: It is common to use a content inventory as the basis of a hierarchical site map. If the URLs represent a logical directory structure, you have a great start at creating that map. If not, a site migration can provide an opportunity to redesign your site's URL structure to better reflect the content hierarchy. Your audit is your opportunity to make the case for addressing the issues in your current structure.

For more detailed instruction on how to construct effective URLs, there is no better source than Google's own Webmaster Tools documentation.[2]

Type

The type, or format, of the content – for example, HTML, video, image – is another basic piece of information to identify the overall structure and content mix of your site. Does your site include a large number of PDFs? You may want to flag those for review and/or incorporation into the site in to be more usable and more readily indexed by search engines. Are there videos and audio files? Given that video and audio media are rarely recreated once developed, they can be especially vulnerable to becoming irrelevant and outdated; be sure to include them in your audit.

---

[2] https://support.google.com/webmasters/answer/76329?hl=en

File Size

File size may interest your website management team, who care about the effect of file size on load time and performance.

Metadata: Title, Keywords, Description

Keywords have declined in importance for SEO. However, the title and description metadata are still important. The title appears in the browser and in search results, so it must be unique and descriptive – get those keywords in there! – without being too long. Best practice is to keep the metadata title under 70 characters.

The metadata description also appears in search engine results, so review it to see whether it represents the content on the page and whether it is engaging or informative enough to entice readers to click through.

Analytics

Analytics data give you a good indication of which pages are popular and whether people are spending time on them or immediately leaving. If you have an analytics program like Google Analytics enabled on your site, export the data for your pages and copy that into your inventory.

Copying and pasting analytics data into an inventory can be a large, tedious, and error-prone exercise. This is where an automated inventory tool can save time and ensure accuracy. See the section titled "The Tool-Supported Inventory" below.

Use the analytics data to find pages that have little or no traffic, and flag those for review; they may be suffering from ROT (redundant, outdated, or trivial content). You may also have an instance of valuable content that's been orphaned without navigation or buried too deeply – another reason to look at the low numbers as well as the high.

Related: Chapter 10, *Auditing for Content Effectiveness*

Word Count

If you're planning a localization project, knowing the word count on each page is a helpful tool for estimating scope and cost. Even if you're not localizing, it's a quick way to find very long or very short pages that you may want to review. Short pages with little content may need further development or you may be able to incorporate that content into another page; long pages may need to be broken up or edited for scannability.

Custom Data

Besides the typical inventory elements, many people gather information that is relevant to the project and team and that sets the stage for the content audit. For example, you may add a column to indicate a status (*revise*, *remove*, *retain*), a business owner name, a step in the customer journey that the content maps to, a persona for whom the content would be appropriate, and so on.

H1 Tags

H1 tags are important for search engine optimization. Review them for keyword placement and clarity.

Links In and Out

A good way to assess the site's cross-linking strategy is to look at the links into and out from each page. You can also use the list of links in to find pages that are not linked from expected locations or that are available only from the sitemap or a low-level page.

Images, Media, Documents

In addition to HTML pages, gather data for the images, media (audio and video files), and documents that appear on your site. Knowing which files are associated with each page helps you plan and track content through a migration process.

Image alt text is important for accessibility compliance. Include alt text in your inventory if possible.

The Tool-Supported Inventory

Depending on the size and complexity of your site, you may find it helpful to have the benefit of an automated tool, whether a file-list export from your CMS, the output of an search engine optimization tool, or the data provided by a tool such as the Content Analysis Tool (CAT),[3] which creates content inventories. If you choose to create your inventory in the Content Analysis Tool, you can quickly amass both summary and detailed site information which you can view as a dashboard or export to Excel.

Creating an inventory in the Content Analysis Tool consists of a simple job setup process in which you tell the tool the URL from which to start a crawl and set other parameters to refine the results. When a crawl has

---

[3] http://www.content-insight.com/products

completed, the job details are presented in a tabular format in an inter-active dashboard. You can sort, filter, and add your own columns of data into this view.

A tool-supported inventory provides a count of content types, which gives a quick sense of the overall size and scope of the site. You may find it handy to filter the results by content type or status, or you may want to find broken pages (indicated by code 404) or *redirects* (indicated by codes 301 or 302). You may also want to view a list of files that you can sort by the URL, type (format), site level of each page, title, word count, or Google Analytics data (see Figure 4.1).

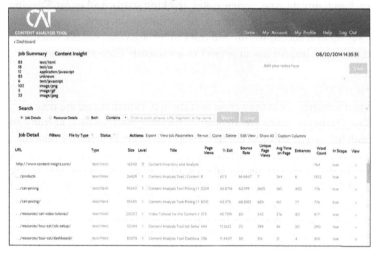

Figure 4.1 – Example of the information available in the Content Analysis Tool

Here are some things that you might want a tool like CAT to enable you to do:

- See the metadata for a given page (title, description, keywords)
- See all the files associated with a given page (images, documents, audio, video)
- View the links in and out from each page to evaluate cross-linking strategy or find pages with minimal navigation
- View and save a screenshot of the page
- Quickly review the metadata against your site standards
- See word count, helpful for estimating localization
- Find pages that may not have any images so that you can consider whether you want to liven them up

- Add your own columns and tag files with values relevant to your project, such as business owner name, page type, or status
- Add your own notes about the page for your own reference or to share with colleagues

Building Out Your Inventory

Once you have gathered your basic data, you may decide to add columns, either by using the Custom Columns feature in CAT or by adding columns to your spreadsheet manually. Over the course of the content inventory and audit process, your inventory may grow to include data such as template type, review notes, tracking status, and so on. Keeping your inventory up to date during the lifetime of the project ensures that all content is accounted for and tracked through the process, whether it is a content improvement project or a complete CMS migration.

Refreshing Your Inventory

Most websites constantly grow and change. Whether you are doing a short-term project or are involved in ongoing maintenance, keeping an updated inventory of all content assets has many benefits for ongoing governance. Making the inventory part of your website lifecycle enables you to always have an up-to-date snapshot of the content you are making available to your users at any moment. Keeping your inventory refreshed also puts you in a good position when the next project comes along – you can easily provide data to help scope a new audit or migration and identify any pages or sections that are due for updating or archiving.

If you use the Content Analysis Tool, you can run another crawl of your site during the project and compare the two jobs to find pages and files that have been added, changed, or deleted. If you are doing a manual inventory or creating reports from your CMS, gather a list of files and updates, and incorporate that information into your inventory.

Turning an Inventory into a Site Map

By sorting and organizing your inventory list, you can create a spreadsheet view of the overall site structure – a *site map*. Order the rows according to the site structure and, if you find it helpful, add a column with a numbering system to indicate page hierarchy. The resulting sheet may not be as user-friendly as a more visual representation, but it can provide a valuable way to communicate the revised site structure to the people who are rebuilding your site.

Turning an Inventory into a Content Matrix

A *content matrix* is another form of enhanced inventory. In *The Language of Content Strategy*, Sarah Beckley defines a content matrix as:

> ...an expansion of the content inventory to track the progress of each piece of content through the stages of a project or content lifecycle.

What I call a content matrix is slightly different. It is a view of the content inventory that represents the future state of the website from a structural perspective (by organizing it into a site map) with the content for each page added in, including the copy (or references to its location) as well as lists of the images and other files that will make up the page. A matrix can serve the production team as a blueprint of the new site.

Summary

The content inventory, a comprehensive accounting of all the content assets on a website, is an important first step in any website project. Understanding exactly what you have at Point A makes plotting the path to Point B easier. Start with the basic data, and build out whatever information is most valuable for helping you evaluate the next steps and make decisions. Inventories give you a starting place by establishing the size of the audit effort, and the data points you gather allow you to identify structural and qualitative issues that you'll dive into more deeply during your audit.

Basic Elements of a Content Inventory

- URL
- Type (HTML page, image, document, audio, video)
- Metadata (title, description, keywords)
- Publication date
- File size
- Analytics data
- Links in and out
- Word count

CHAPTER 5
Preparing for a Content Audit

Content is at the heart of a unified content strategy. Before you can model your content – and subsequently, unify it – you need to gain an intimate understanding of its nature and structure.
—Ann Rockley and Charles Cooper, *Managing Enterprise Content*

In This Chapter

After you've completed your content inventory, you're ready to audit your content. Deciding what and when to audit, establishing the scope, and immersing yourself in the business context are critical steps in planning your audit process.

Turning an Inventory into an Audit

A content inventory provides you with the baseline and current state of the site, providing the template for adding the information that turns an inventory into an audit. Gather and review all the data that's available about the site and your customers, and establish your scope.

Chapter 4 looks at how you can work either within the Content Analysis Tool dashboard or in a spreadsheet template of your choice to build out the structure of the content audit. To the data gathered by the tool, if used, add columns for the other information you wish to capture for each page, such as content type, content owner, status, and notes. This becomes an ever richer set of data as you evaluate all those pages.

In deciding how to analyze content from your audit, consider your goals, audiences, timeline, and budget. Depending on your goals, you may choose an audit rubric such as Rahel Bailie's *RAITES method* (relevant, accurate, informative, timely, engaging, standards-based) to add additional evaluative criteria, or you may create your own criteria.

Determining What to Audit

A thorough content audit assesses content, structure, and functionality affecting external users – and issues affecting internal teams who create and manage the content. Depending on your audit goals and scope, you may assess all or some of the following criteria about all content types: textual content, media content, and images.

External issues:

- Content quality
- Content effectiveness
- Relevance
- Messaging
- Branding
- Language support
- Structure of site and content
- Navigational model
- Interaction models

Internal issues:

- Content consistency where appropriate for reuse
- Individuality where needed for discrete markets or audiences
- Terminology
- File-naming conventions
- Asset management
- Templates and page types
- Content types
- Content workflow and lifecycle

A complete audit can encompass any or all of these factors. Audits, ideally, are a multidisciplinary effort. Some aspects, such as the structural, are audited by information architects and business analysts; content strategists and content managers look at the content quality, breadth and depth, and interconnections; site optimization people look at effectiveness as measured by analytics and other metrics; and the marketing team and the creative team look at images, messaging, and brand support. (For more on who should participate in an audit, see Chapter 3.)

When you're building out your audit, starting with your inventory data, you add this kind of information so that ultimately you have a complete picture of all your content – where and how it's managed, what's translated and what isn't, what is on brand and what isn't, and so on.

In later chapters, we discuss audit methodologies or areas of focus. Before you begin, it's helpful to prepare by establishing your scope and familiarizing yourself with the resources you are auditing against.

Scoping the Audit

Chapter 1, *Building the Business Case*, discusses setting audit goals. Closely aligned with establishing goals for your audit is the task of setting your scope. Goals are about the reasons for auditing; scope is about how much auditing you're going to do, what you'll focus on, and how you'll know when to stop.

Depending on your project context, goals, and timeline, the scope of your audit could range from a high-level overview of top pages and representative page types to a detailed, page-by-page deep dive.

In her blog post "How To Conduct a Content Audit,"[1] Donna Spencer offers this advice:

> Don't capture information you are unlikely to need or use. If you're unsure whether you need information for a specific page, write it down for a handful of pages, to get a feel for whether it will be useful. You can always come back and fill it in for other pages at a later stage.

How do you know what information you are likely to need or use? The following sections help you answer that question.

Your Time

If you have limited time in which to conduct your audit, constrain your scope and focus on the content most directly related to your goals. Chapter 1, *Building the Business Case*, discusses how your goals may be affected by your audience, the questions you need to answer, and the project timeline. If you are auditing to scope a project, for example, you may be focused mainly on the quantitative aspects to establish the size of the project.

As discussed in Chapter 10, *Auditing for Content Effectiveness*, you can also use your *analytics* data to help determine what to focus on. For example, you might apply the 80/20 rule and focus your attention on the top 20% of visited pages, or look at the pages you've designated as most important for *conversion*.

If your business goal is to improve a specific user journey through your site – a purchase buy-flow, for example – you may take a vertical approach. Focus on the content that supports that journey and follow it through the entire intended path. If you have site traffic data that indicates users are following a path other than the one you intend, that can be valuable information to consider in your audit.

If your site has many similar pages – for example, an e-commerce site with thousands of product pages – you may be able to sample a cross-section of those pages. Assess one representative page from each product

---

[1] http://uxmastery.com/how-to-conduct-a-content-audit/

category. If you aren't reading each product description for language or other specifics, that sample may be sufficient to find issues or draw conclusions.

If your goal is to assess the site as a whole but your time is too limited to accomplish this, you might take a horizontal view, looking at all top-level and second-level pages, or as deep as you have time to go.

Project Type

There are several scenarios in which you might perform content inventories and audits:

- Content strategy initiative
- Website redesign
- CMS implementation
- Global market rollout
- Governance initiative
- Ongoing maintenance

The type of project you are initiating can influence the scope and focus of your efforts. If your audit is for the purpose of providing scoping information that allows someone to make decisions about the viability or cost of a project based on the information you're gathering, you might focus on the sheer numbers – how many pages, how many images, how many words, how many videos, how many interactive features (logins, forms, transactional elements, and so on).

A content strategy initiative implies a deep assessment of content quality. You would look at relevance, currency, writing style, breadth and depth, performance, and how well the content supports business and user goals. Your audit scope will probably be both broad and deep to ensure a thorough enough evaluation to support the initiative.

In a site redesign project, the goal of the audit is to evaluate the content against a new design and information architecture. If there is no business need to carefully review each piece of content for quality or accuracy, the audit scope can be limited to identifying representative pieces of each type of content on the site.

For example, your audit might examine the home page, a first-level landing page or category page, a second-level landing page or category page, an article, a press release, a blog post, a company information page, an executive biography, and so on. If you're working on an e-commerce site, you'll want to look at a product-gallery page, a product-detail page, and campaign or promotional pages.

If the site is being migrated to a new CMS, your scope is larger. In this case, you must review each page individually, particularly if you are weeding out old, inaccurate, or substandard content. While it is always good practice to share the content review task among the business owners or stakeholders, inclusiveness is even more important when you are looking at the entire site. Not only does breaking up the work help make a large task more manageable, it enables the subject matter experts to make decisions about what to do with their content. It also enables working in parallel, which is helpful if your timeline is tight.

For a *globalization* project, focus on images, word count, language issues (such as colloquialisms that present translation issues), and the core set of content targeted for translation.

If you have adopted the concept of the rolling content audit for ongoing site maintenance, you can scope your audit to only those pages that are new or have been recently changed and to regularly reviewing old content. (Some CMSes enable you to set a review flag to remind you to review at a set interval.)

Related: Chapter 9, *Auditing for Content Quality*; Chapter 10, *Auditing for Content Effectiveness*; Chapter 12, *Auditing for Global Issues*

Business Context

Chances are, if you're working on a site audit, you are doing it for the sake of a larger business goal. Maybe a site migration is planned, or there are new business initiatives related to the site, or there is a general sense that it's time to give the content a good once-over. Your audit can play a role in communicating to your stakeholders on both the business and technical sides of the house.

Stakeholders are not only the audience for the audit, they can also provide necessary input about scope. Knowing the limits of the organization's budget and appetite for change can help set expectations and avoid time lost to gathering information that isn't relevant to the business goals. If a business decision is made, for example, to migrate all content as-is, time used to evaluate how content should be rewritten is not time well spent – whether or not you agree with the decision!

Later, we'll talk about the value of doing a rolling audit for the sake of ongoing quality control.

Project Timeframe

Another lens through which to view your scope is the length of time you'll be using the data and the rate of change over that period.

If you're doing a site migration project, depending on the size of the site, you may work for a year or more, during which time the site will continue to evolve and grow. It might not make sense to do an exhaustive assessment of content that may not exist by the time you migrate. How much will the content change over that period? Does it make sense to catalog and evaluate every piece of content, or is it enough to gain a general sense of scope?

What kind of site is it? If it's an e-commerce site, it changes frequently, and all product content is probably database-driven. So you don't want to spend a lot of time capturing and reviewing all those pages since they're going to change and since you risk skewing your scoping efforts if you count each product page as a separate page to be migrated and rebuilt.

A long-term project is also a case for a rolling inventory. You need to revisit the inventory periodically throughout the project to ensure that your scope is still accurate and that all new content types or templates are accounted for.

Project Limits

Although you may be doing an audit as the groundwork for a project proposal, keep in mind your overall business context and resource constraints. If you're in a small organization with limited resources, spending weeks or months identifying issues that you won't be able to fix isn't the best use of your time and won't impress your management or client. If you know that can't do a complete site overhaul or that, at best, going big will be a hard sell, start small. Focus on the issues that can be easily identified and fixed so that you can get some quick wins and earn the trust and respect of your stakeholders, who may then be open to a larger project.

Setting the Stage

Before diving into an audit, assemble and familiarize yourself with as much of the business context as possible. Time spent clarifying your goals, scoping the audit, and immersing yourself in the business context helps you spend time wisely and achieve the results you need.

Business Requirements

The business requirements, usually gathered at the start of a project, tell you the expectations and what the business wants to achieve with the content. If you are working in-house or outside the context of a dedicated project, you may have access to the established standards that the business uses to measure success.

Your stakeholders are a vital source of information about the business goals. They are also a great way to learn about the pain points and the silos that you may need to break down. There is no substitute for hearing directly from the people who own or manage the content. Ask them for their pain points and their wish lists, and determine how the current workflow and the site content play a role.

Other sources of business goals include the company's mission statement, business strategy documents, or roadmap. If you don't know what the goals are for the site and how they're measured, you don't have an accurate baseline.

Site Data

Assemble all your sources of site performance data – traffic, page views, conversions, and so on. This is important if you plan to audit for performance issues like low traffic or pages that don't convert.

Brand and Style Guidelines

Before you begin your audit, gain an understanding of what your brand is intended to represent and how it is supposed to be expressed. You'll be looking at content and images in that light.

Corporate brand guidelines usually address rules for content creation, including how to refer to the company's name, trademarks and registration marks, usage of taglines, and voice.

Brand guidelines also usually address location and placement of logos, acceptable colors, icons, and other design elements. This part of your audit may be done by your marketing or creative teams. If you're working on a site redesign or globalization project, pay extra attention to reviewing these elements.

Customer Research and Data

Just as we need to understand the business owners and their requirements, we need to understand the users and their requirements.

If *personas* or other customer research were created as part of the project, get familiar with them. Data-driven personas provide a useful lens for viewing the content and testing it against what you know about your customers. (For more information on how to use personas, see Chapter 8, *Using Personas and Customer Journeys in Audits*.)

An even more powerful approach is combining personas with *customer journey maps*. If your site has a definable path or set of tasks that users typically go through – or that you want them to go through – create a customer journey map, and associate content with each step. This enables you to find gaps – places where you aren't supporting your customers enough or where support levels are inconsistent.

Maybe you even have too much content at certain steps, more than your customers need. In that case, take a look at the analytics for those pages to see which content is being used and whether you can reduce it to keep only what's most useful. (For a sample, see Appendix E, *Customer Journey Map*.)

If you don't have personas, learn everything else you can about your audience. If you don't have or can't afford primary customer research, learn what you can from secondary research sources such as industry analyses and trends.

Mine carefully any data that comes directly from your customers. If they leave feedback in the form of reviews or ratings, or if they post questions to a Q&A or FAQ section, you have a good idea of how engaged they are with that content. And if your customer service team is getting a lot of the same questions over and over, look to see how or whether you are answering those on the site. You may need to create content to help answer those questions, or maybe you have the content but it's too hard to find. Either way, you can take action based on that information.

Once you've immersed yourself in all of this information, you're ready to dive into the content itself if you haven't already.

Summary

Building on the foundation of your content inventory, establish what to audit and when. Consider the larger project context, and determine your audit focus and scope.

Determine Audit Criteria

- Content quality
- Content effectiveness
- Relevance
- Images
- Messaging
- Branding
- Language support
- Structure of site and content
- Navigational model
- Interaction models
- Content consistency
- Terminology
- File-naming conventions
- Asset management
- Templates and page types
- Content types
- Content workflow and lifecycle

Assemble Audit Resources

- Inventory data
- Business requirements
- Analytics data and other metrics
- Editorial and brand guidelines
- Personas
- Customer journey maps
- Customer feedback
- Search logs

CHAPTER 6
Ready, Set, Audit!

If you have read this far, you have learned that content inventories and audits are the building blocks of a content analysis project. We begin with these activities because they establish a baseline from which we can measure how far we've traveled when we turn our insights into actions. The data gathered in an inventory enables the auditor to establish scope, get a sense of the structure, and identify patterns in the current content. The inventory format, whether in a tool or spreadsheet, provides the structure for information to be added as the audit proceeds.

Inventories and audits can be done for a variety of reasons, but they always need to be placed in the larger business context. This not only helps the project team have a shared understanding of the goals and scope but also ensures organizational buy-in. Inventory and audit projects can be time- and resource-intensive, and you may need to convince the people writing the checks – whether external clients or internal managers – to spend the money and take the time. Applying project management rigor to an audit project is a good way to show the check writers that this is a well-managed task with defined outcomes. It also helps ensure that team members understand their roles and responsibilities.

We've talked about the what and why. Now let's talk about the how.

Content can be audited from a number of perspectives and, depending on the time, resources, and context, may include any or all of the methodologies discussed in the following chapters.

The most basic audit is the one I call the *qualitative audit* (as defined in Chapter 9, *Auditing for Content Quality*). Of course all audits are qualitative, but this audit is focused on traditional editorial qualities, such as relevance to the audience, currency, consistency, and usefulness.

In Chapter 10, *Auditing for Content Effectiveness*, you will see how to overlay analytics data as a way to assess how content is being used.

Just as we audit our own sites, we can also audit competitor sites (as described in Chapter 11, *Auditing Competitor Sites*) and do a side-by-side comparison, focusing on key areas in which we want to compete.

Chapter 12, *Auditing for Global Issues*, looks at the issues, such as terminology and imagery, that can trip us up as we globalize our sites.

Chapter 13, *Auditing for Legal or Regulatory Issues*, outlines some of the content-related issues we must be aware of to protect our organizations from harm.

Finally, once we have all our data gathered, our content assessed, our analyses completed, and our strategies for improvement developed, we need to share them. In Chapter 14, *Presenting Audit Findings*, you will find advice about choosing which data to present, based on your audience, and how to present it effectively.

Ready to audit? Get your inventory, open your browser, pour a fresh cup of coffee, and dive in!

Building and Delivering the Audit

Content audits can be focused broadly or narrowly and include a number of different tactics. Depending on your business context and project goals, you may choose to focus on issues such as content quality, competitive advantage, performance, global issues, and legal issues. In this section, learn how to audit for those issues and how to incorporate regular inventories and audits into your governance process. When your audit is complete, present the findings in a compelling way that make it easy to take the next steps.

CHAPTER 7
The Multichannel Audit

In This Chapter

A comprehensive content audit looks across all the channels to which a company publishes content, including digital and print sources, to ensure consistent quality and messaging.

Auditing across Channels

It is a rare company these days that produces content for a single channel. In our always-connected societies, people engage with organizations, brands, and content across multiple touchpoints. In addition to a website, a company may also have print materials such as catalogs and brochures, other digital communications such as email or mobile applications, and of course, social media. Retailers may also have in-store experiences such as signage or interactive kiosks. This content ecosystem needs regular tending to remain healthy.

Multichannel publishing brings many benefits to a business, supporting a customer journey through the steps from awareness to conversion and ongoing engagement. It also carries with it the risk that content can be overlooked, fall out of sync with other content, or become off-message and off-brand.

Just as a successful content strategy accounts for all the contexts – and even devices – in which a customer encounters a business's information and messaging, so too does a content audit.

Keeping track of content and auditing it across channels is easier if it is authored and managed in a centralized CMS. Lacking a CMS, the auditor's job is to search out all those channels and audit them individually, keeping in mind the goals and standards that the business strives to achieve.

Where Do We Look?

Begin by identifying all of the channels to which the company publishes. Survey the marketing department, the product group, the sales team, the human resources team, the media-relations people – anyone who might be responsible for content creation and publication. While you're compiling your list, be sure to ask the content owners to help you understand the goals and purpose of each piece of content created, so you'll know how to assess it individually and against the other content types.

In your audit spreadsheet, you might document data points like these:

- Channel name
- Format
- Content types published
- Frequency of publication
- Last-published date
- Primary audience
- Content owners or creators

You might also include columns for notes on voice and tone, content style and quality, and any other evaluative comments.

How to Audit Multichannel Content

What do we audit for across multiple publishing channels? The same things we would look at in a single channel – quality of writing, consistency, accuracy, currency, voice, and tone. Wherever a customer engages with your content, he or she should feel that it all came from one company. We also need to look at the nuances of each channel and the ways those criteria might vary depending on the source.

Key to auditing content in any channel is understanding the audience. Product content produced as in-house material for a sales team is probably written differently than for a customer catalog. But certain aspects of that content, such as product specifications, must be consistent. A newsletter may take a different tone in telling the same story as a press release. Content produced for social channels may have as its goal dissemination of information about the company, but does it differently than a company backgrounder that is prepared for investors. Presentation, format, content length, voice, and tone may vary across these channels, but factual information about the company and its products and services should not vary.

As the focus of this book is on auditing digital content, I won't delve further into print other than to reiterate the point that print sources are even more likely to become outdated or incorrect. And unlike a website or social media post, print sources can't be retracted easily once they are out in the world. So it becomes even more important to include print content as part of your ongoing audit and governance processes.

Related: Chapter 9, *Auditing for Content Quality*; Chapter 10, *Auditing for Content Effectiveness*; Chapter 11, *Auditing Competitor Sites*

Auditing Social Media Channels

With the proliferation of social media outlets comes a corresponding uptick in their adoption by companies looking to reach their customers at every possible point. The rush to publish to Facebook, Twitter, G+, Instagram, Flickr, Pinterest, and more has resulted in companies hastily creating online presences that may not be fully in line with business and content strategy. Not only are there multiple company identities out there, but they may be managed by separate people or teams within the organization, and they may not be presenting a consistent, on-brand face to the world.

Include social media research as part of your audit to ensure that you're getting the most complete picture possible of content being produced by your company.

Finding all the channels to begin with may present a challenge unless your company has carefully managed its social presence. In addition to the official list of profiles you can gather from your stakeholders, try searching on each site for your company name or keywords associated with your business just to make sure there aren't rogue sites out there that aren't being managed. Once you have your list of channels, you can begin to gather information about them.

A good list of data to start building your audit includes these:

- Network name
- URL of your page or profile
- Type of content being shared
- Any visible metrics (likes, shares, followers)
- Date of most recent update
- Audience
- Business owner

To your audit, add your evaluative criteria:

- Consistency of information, such as company profile
- Broken or outdated links
- Style and tone of posts
- Branding
- Performance metrics

You also want to look at how well or poorly the social network's presence is integrated with the other publication channels. Do the profiles or accounts link back to the website? Does the website encourage users to follow the company on the networks? Are you pulling in social content such as the Twitter feed to the site? Is content cross-posted between the blog and the social sites?

If resources for managing social networks are limited, assess how well each channel is doing, via hard metrics such as analytics data and soft metrics such as sentiment analysis (how favorably or unfavorably your company is being spoken about). This helps you focus your efforts on the channels that are most effective for your goals and your audience.

Most social media channels have their own metrics (Facebook Insights, for example), but you can also employ third-party tools that track social engagement statistics. Google Analytics shows what traffic comes to your site from each channel and what content is visited most.

Look at other engagement metrics as well, such as whether questions or polls are being answered, whether comments are being posted on blogs, whether comments or reviews are positive or negative, and so on.

Like any other aspect of your content audit, you can also incorporate a *competitive audit* into your process. Look at the social presence of your top competitors and see how well they do on the same metrics of engagement you track for your own. If you share a target audience and your competitors are scoring higher with them, you can use that comparative data to reset your own baseline and goals.

Summary

Publishing content across multiple channels, including the wide variety of social media networks, gives a business an expanded reach and additional opportunities to engage customers. But managing content across all those channels presents challenges. Regular content audits that include all channels play a key role in content governance, ensuring a consistent identity and high-quality information presentation.

CHAPTER 8
Using Personas and Customer Journeys in Audits

In This Chapter

Personas, especially as an extension of efforts to segment your target audience, are an important tool for content strategists, user experience architects, creative directors, product planners, and marketers of all kinds. Personas provide a useful, agreed-upon starting point for nearly all forms of marketing, digital or otherwise. They also provide a great opportunity for content strategists to align their work to a business's larger digital marketing strategy. Combine personas with *customer journey maps* for even more insight into your customers' needs.

What Does a Persona Look Like?

Auditing website content against personas can result in a tactical set of actions and content plans. Let's look at what personas are, how to develop them, what information to add to a standard inventory and audit, and how to extract insights and action plans.

What is a persona? A simple description, from Usability.gov, says:

> A persona is a fictional person who represents a major user group for your site.

There are different methods for creating personas. Most include at least these basic facets:

- Personal data such as name, a photograph, family circumstances (married with children, single, etc.)
- Geographic location
- Demographics such as age, occupation, income, and if relevant, ethnicity

To make a persona meaningful as a way of answering experience and content questions, a persona often includes "softer" data such as needs, motivations, and behaviors, along with social technographics and media-consumption patterns.

Personas can be simple or complex, and great resources exist for learning how to create them.

Related: Appendix D, *Sample Persona*; Additional Reading

Developing Personas – A Collaborative Approach

> Because so many different departments within an organization have a stake in the content that is produced, the single-most important investment a business can make is to commit to researching and creating personas that will be used to make all content decisions going forward.[1]
> —Kris Mausser, "Personas: A Critical Investment in Content Strategy"

To ensure that these valuable tools are available to us as we approach the content strategy, we need to make sure we're involved in the user research and in the process of persona creation. Personas are often developed by the user experience architect who may well have an awareness of the importance of the content, but who is still likely to be primarily focused on interactions, navigation, and functionality. If we can partner with our colleagues on mapping the overall customer journey, from the functional perspective and with a view to the content needed to support the customer along each step of that journey, we can build richer, more actionable personas.

It is critical that personas are built on real data, not on assumptions about customer types and their behaviors. Contextual interviews with actual customers are often a starting point; additional data may come from third-party research. (For example, if you're working on a site for a travel company, you probably have ample research data about the demographics and behaviors of travelers.)

Once the personas have been developed, advocate for collaboration with the larger team, and perhaps business stakeholders, when it comes to applying the personas. The goal is to establish a common understanding of the personas and how to use them to create the audit categories and, eventually, the site content itself.

One valuable technique is to conduct a brainstorming session where team members advocate for one of the personas. After brainstorming, then you can deduplicate, test, refine and prioritize your audit activities. Keeping the personas alive, so to speak, throughout the project can provide a way to make sure the content and design stay on track. One way to keep the personas in your awareness is to print the persona documents and pin them on your office wall, a hallway, or any other public space frequented by the team.

---

[1] http://discontentedcompany.com/2012/10/27/personas-a-critical-investment-for-content-strategy/

Periodically revisit and update your personas over time (even after your current project has launched). As new content and site features are created, check back against your core personas. Use the data you gather about your customers to continue to refine your personas and your personas will continue to serve you well.

Communicating Personas

Personas facilitate cross-team understanding of your user goals and can provide a common language for the project team. Well understood, well-communicated personas and user scenarios or customer journeys can help the team agree on priorities and implementation strategies. You may discover in the process of developing your personas that existing content does not support their needs. This information contributes to your content gap analysis and provides a clear framework for content development goals. Quantify each persona's need for a type of content, assess whether your content is delivering what your personas need, and make sure your content plan addresses the gap.

Personas can also help make the business case for your content project. If your personas are clearly based on targeted audiences and you tie the content to scenarios such as awareness/consideration/conversion paths for them, you can make the case for investment in improved content.

Adding Content Strategy to Personas

In "Include Your Clients in the Persona Research Process with Affinity Mapping," [2] Norris A.A. Rowley, Jr. writes:

> Persona research is central to the content strategy process because it allows us to create content that speaks directly to the users we want to target. We are able to gear our messaging toward the interests, aspirations, concerns, and desires of the very people we've identified as being crucial to our client.

What would a content-focused persona look like, and how would we use its data to audit existing content and create a strategy?

---

[2] http://www.iacquire.com/blog/include-your-clients-in-the-persona-research-process-with-affinity-mapping/

Getting Started

A good place to start is by listing these things:

- The user's potential questions
- The user's information needs or level of awareness/expertise in the subject matter
- The user's potential problems (not limited to the product or service)
- The user's potential paths/tasks (the customer journey)

Measuring against Goals

Developing a set of heuristics for each of these goals makes them measurable and actionable. Here's what to look for:

- Content written for at least one clear audience, including tone and voice nuances
- Content reflecting that persona's interests and information needs
- A call to action within the text (not leaving it up to navigation)
- Content mapped to a step in the customer task flow

Adding Persona Data to Your Audit

Starting with your base audit, add columns for the data you've decided to track. For example, you may add other elements like these:

- Persona name (or type)
- Customer task (evaluate, learn, purchase, get support, share)
- Call to action
- Status (content is okay as-is or needs revision)

Add lines to your audit for new content to be created and, as discussed above, review your findings with your team members and stakeholders.

Applying Insights to Action

Evaluating content against these heuristics and including analytics can focus your content strategy for maximum impact.

Defining calls to action that answer the short list of questions/problems/needs of a persona saves valuable time in scoping your audit. Without this framework, you can throw any kind of user need at the audit and not know where to stop. Personas tell you what the priority is and how deep to go. Some customers are worth 80% of your time, some are worth 20%. Apply that metric to how much time you spend on the audit.

Even defining content types can be a persona-related question. For example, instructional content is very different from product page content, which is very different from sales content. Personas can help determine how much of each kind you need and how it should be structured. For example, customers need different content in the presales evaluation/awareness step of their journey than they do in a postsales support step. Work closely with the user experience designer to make sure that there are clearly marked entry points to these content types and then make sure that once your persona reaches the destination, the content is appropriately written, offers the right kind of information, and helps them know exactly what step to take next.

Landing pages play a big role as well since they are often created for target audiences or segments. Analytics help you know which ones are working – and even where that persona is located and what type of device he or she used to access the content. This also applies to gap analysis; through your audit, you may discover there are currently no landing pages that serve a particular persona's needs. By using a persona-oriented audit, you'll be in a great position to recommend adding the relevant material.

When you've completed the audit exercise for your highest-value content, you have the foundation for your action plan and an informed content strategy. Ongoing testing of your decisions and assumptions against real-world data, using your analytics, customer feedback, and other sources of user input helps you continuously refine your site content.

Customer Journey Maps

On their own, personas are a great tool for helping ground your content strategy and user experience in the real-world needs of customers. Combining personas with customer journey maps offers yet another level of understanding.

A customer journey map captures a user's step-by-step experience as he or she interacts with your content, at all touchpoints – for example, a retail customer-journey map may begin with the customer seeing an advertisement on television, going to a website to do more research, and ultimately making a purchase in a physical store. The journey map for that customer documents the content and experiences at each point and describes his or her state of mind and what information is needed to take the desired next step at each point. Mapping out a current journey versus an optimized journey can help show gaps in current content and point to areas for improvement.

Create a separate content map for each persona to ensure that you're covering all your content and representing all your user groups. Use the quantitative data you have on your customers, from user research and analytics, as well as the qualitative measures such as the depth, style, and tone of the content that supports each step.

Customer journey maps can take many forms, from simple tables to elaborate diagrams, but if you don't need to create a polished deliverable, use whatever format feels most intuitive to you and your audience. Creating customer journey maps can be a great workshop activity, bringing stakeholders together to do a collaborative review. A whiteboard or sticky notes and easel pads are great tools for low-fidelity mapping. By their nature, the maps are visual representations that can be quickly and easily grasped, making them great additions to an audit presentation.

Related: Appendix E, *Customer Journey Map*

Gap Analysis

The personas and customer journey map can be further developed to support *gap analysis*, or the process of assessing the difference between current state and future state. Using a form, such as the Gap Map in Appendix F, allows you to combine what you know about your users and what content will support each step of their journey through your content, from initial awareness to conversion to final loyalty and advocacy of your brand.

Start by aligning users and their journey on a grid (see Figure 8.1).

Audience	Buyers' Journey			
	Awareness	Consideration	Conversion	Loyalty/Advocacy
IT Professional				
Decision Maker/C-Suite				
Influencer				
Audience	User Journey			
	Awareness	Consideration	Apply	Recommend
Career Seeker				

Figure 8.1 – Gap analysis grid

Complete the exercise by adding user goals and business goals for each step of the journey. An example user goal might be "Help me understand the value of your product or solution." A sample business goal might be "Drive traffic to product or solution content." The content required, therefore, might be an engaging product video prominently placed on the home page. Mapping out the needs and content gaps helps align your most important business goals with the user goals when you are trying to evaluate whether a piece of content is on strategy or not.

Summary

User personas offer a valuable framework against which content can be measured. Creating rich personas based on actual user data and aligning them with business goals can help make the business case for investment in content and can facilitate creation of very targeted content improvement and development plans. Incorporating personas into customer journey mapping provides a helpful illustration of how and where users interact with your content and where the strengths and weaknesses lie. This cross-channel review is a great addition to your multichannel audit.

CHAPTER 9
Auditing for Content Quality

In This Chapter

A content audit is usually defined as the qualitative view of your content. While data, such as analytics, can be useful in evaluating quality, audits usually involve the more traditionally editorial measures, such as assessing against brand guidelines, style guides, and voice and tone. In this chapter we'll look at some of those qualitative measures and ways to evaluate content against them.

The Qualitative Audit

What makes a website good? We often say that content strategy is about getting the right content to the right people at the right time. "Right content" in that phrase is usually taken to mean relevant, timely, useful – all good things. But what about content quality?

Every time a user comes to your website, you are given an opportunity to convey information, inspire trust and loyalty to the brand, make a sale, or otherwise engage the user. How you speak to your customers, whether your content seems thorough and credible, and how consistent the experience is compared with other ways your customers engage with you, such as your print materials or retail presence, can have a significant effect. The qualitative audit entails evaluating content against a set of editorial measures.

Related: Appendix C, *Content Audit Checklist*

What to Assess

A standard set of audit criteria includes these attributes:

- Content is relevant
- Content is current
- Content is accurate
- Content is engaging
- Content is easy-to-read/scan
- Content is audience-appropriate
- Content is consistent
- Content communicates key messaging
- Content facilitates key user activities
- Content has sufficient breadth and depth
- Content is appropriately formatted

Let's look at how to assess these criteria.

Relevant

How do we know if content is relevant? We ask ourselves if it serves a business purpose. It may be a beautifully written article or a fun video, but if it isn't related to the core business purpose or a customer task, it may not be appropriate for your site. Your users will also tell you what's relevant and what isn't. Look at the data. If a piece of content is seldom accessed, or users don't spend any more time on it than it takes to click away, it may not be worth keeping.

Caution: the content may be just fine – and appropriate for users – but may not be for the users who found it in that location. It may be great content for someone in the evaluation stage of a purchase journey, but you've placed it at the post-purchase support step. Or it may be misplaced or labeled or written in a way that misleads, and the intended audience isn't finding it – but an unintended audience is finding it and is unhappy. Maybe it looks as though it's going to be introductory, 101-level content, but is actually written for a highly technical audience. This is an example of why you can't always trust the data or any single measure. It's always good to cross-check against other measures, like the business goals or the customer journey.

Current

Content currency is a fairly straightforward one. How old is the content? Does it reflect any updated brand style guidelines in terms of how it's written? Or does it seems out of place with the newer content? If so, mark it for revision.

Accurate

Accuracy, too, has a fairly simple test. Is it still true? This is an especially important measure for product descriptions or support content.

Engaging

Engaging content is easy and fun (or at least not unpleasant) to read. This is a subjective measure, of course, and maybe not even entirely appropriate for your site. If your site is a collection of government policies or tax law documentation, for example, no one will expect it to be fun to read. In general, though, your content should engage your customers enough that they will take the next action you want them to take – buy your product, share the content, learn something, and so on.

Easy to Read

Easy to read content is scannable; it isn't a big, unbroken block of text; and it has illustrations or other visuals if appropriate. To assess, look at your traffic data – how long are people spending on your pages? That may be a clue – if it's a 5,000 word article and contains great information that you want your customers to have and they're only staying on the page for a few seconds, it may be because it's just too hard to read. Mark it for revisions.

Audience-Appropriate

Most websites serve multiple audiences, and each of these audiences needs to feel that the content is relevant to them and approachable. And they need to come away from the site with whatever they came to get. Content strategists need to be as familiar with audience research as the marketing team or any other team in the organization. If you're addressing multiple audiences across multiple cultures, the task of serving up relevant content to each audience becomes even larger. Ideally, each audience needs to feel that the content is relevant and approachable. At minimum, you want to avoid alienating or angering any target audience.

When you understand who you are speaking to, how they like to be addressed, and what they are coming to the site to do, you can effectively evaluate how well the site content meets their needs and speaks their language. For example, identifying key words and phrases that your customers use gives you a clear set of terms to look for in content, links, and headings. Gerry McGovern calls these "customer care words,"[1] and has written extensively about how to mine for and use them.

Chapter 8, *Using Personas and Customer Journeys in Audits*, discusses how to use personas in content audits. Now is a good time to revisit those personas and think about how they would want to be addressed. Personas provide additional angles from which to evaluate content.

Consistent

Content consistency can be measured by looking at the terminology, tone, breadth, and depth of content across the site – does it all seem to have come from the same company or does each article or section of the site feel completely different? Are terms used in the same way? Are products and services described in the same way? Refer back to your brand guidelines and editorial style guides to see how well your content is representing your actual intent and company personality.

---

[1] http://www.customercarewords.com/what-it-is.html

On-Message

Communicating key messages relates to your brand voice and direction, as well as how clear your calls to action are. If you want a customer to take the next step, how obvious is that from the text? If your brand message is something like "We are the definitive source of information for this audience on this topic," does the tone support that? Is the content authoritative and comprehensive?

Supportive of User Tasks

Users come to a website for a reason: they need information or want to be entertained or are there to make a purchase. Mapping the key user tasks into a customer journey and checking your content against that map gives you a quick and easy way to see where content is or is not facilitating user tasks.

If yours is an e-commerce site but it's difficult for a customer to find the information that enables them to make a purchase decision, you may have a content issue to address. You can check this by looking at the path users take through your site, using your site analytics.

Note that the reason a user doesn't find the information may not mean that the content isn't sufficient or appropriate – it may mean that it's in a location on the site that doesn't match the way users navigate. Your action may therefore be to surface the content in a different place or ensure that it's appropriately linked. It may also mean that navigation labels or other terminology doesn't match the way users think.

Sufficiently Broad and Deep

An important metric against which to assess a site is how well the content serves the intended purpose. One way to assess that is to look at the overall breadth and depth of the content. Specifically, is there enough of it? Too much? Does it skim the surface or go into detail? Is coverage equivalent for similar topics? It is also a good idea to review the fitness of content types to information presented.

Breadth

The range of subjects covered will be determined by the type of site, of course, but here is an opportunity to review your personas, analytics, search logs, and business goals. By walking through the customer journeys for your personas, you may be able to determine whether there is additional content that would be useful for your audience. Does the site provide everything this persona would need to know to make a purchase decision, for example, or does s/he need to jump off your site to go learn more first? If so, you've potentially lost a customer. Use this gap analysis to create a list of content to be created.

Conversely, if there is content that is rarely accessed, based on your analysis of site traffic, it could mean that it is either not relevant for your customers or not the type of content they expect to find on your site. It may also be an indication that the content is misplaced on the site or buried too deeply and the problem is with the navigation, not content. This content should be at least reviewed, and maybe targeted for removal or placement elsewhere on the site.

Depth

Look at the content associated with equivalent content types – for example, if yours is an e-commerce site, compare product detail content. How do the product descriptions compare? Are they of equivalent length, specificity, and quality? Do all products (or all within a relevant product category) have a list of specifications? Are the pictures and videos similar in format and quality? If a customer wants to get more information – for example, learn how to use the product – are there links to related how-to or support articles?

In addition to assessing depth by content type, look at depth of content by audience. If your site serves more than one audience, a simple quantitative assessment of how much content is created for each audience may reveal issues. For example, if the quantity of content for equivalent audiences is not consistent, that may indicate that organizational resources aren't being distributed effectively – and is often a side effect of organizational silos, where one business group has more resources or power. Focusing on the user needs and business goals, and being able to demonstrate how and where the site is not serving them, can be an effective tool for creating change. A simple graphic, such as a pie chart, that illustrates content by audience, can be a compelling addition to your audit document.

Appropriately Presented

Content can be presented in multiple formats (text, video, audio) and presentations (blog post, article, support topic, etc.). In addition to all the other lenses through which to evaluate content, take a look at whether the format is the best available for that content type. For example, if users of your site don't tend to watch videos, you may not want to use that format as the only means to present important content about your product or service. Blog posts are expected to be short and ephemeral; they should probably not be used for content that you plan to keep updated or want to be prominent on the site for more than a few days.

Related: Appendix G, *Content Audit Template*

What We Assess Against

Brand Guidelines

Most major organizations have a set of brand guidelines, but unless a fairly robust governance process is in place, it's easy for content creators to either be unaware of them or become lax in following them over time. Before you begin your audit, get access to those corporate brand guidelines, which usually address rules for content creation, including how to refer to the company's name. For example, are registration or trademarks required? Can the company name be abbreviated? Must the company name precede any product name? Brand guidelines may also include usage of taglines and define your brand's voice.

Brand guidelines also are typically concerned with graphics and logo usage as well – for that part of the audit, it is good practice to pair up with your colleagues on the creative team.

Voice and Tone Guidelines

Most brand guidelines will include a set of brand voice attributes. For example, the list may include adjectives like friendly, conversational, upbeat, playful, and energetic for a consumer-facing site; for a technical site, they may be more like authoritative, serious, clear, and respectful.

These are intended to guide the writers in creating content that reflects the brand and sets the tone for the user's experience with the content. When auditing content against these voice and tone guidelines, keep in mind that not every piece of content can or should convey them all and different content types may lend themselves to particular writing styles.

It might not be appropriate to convey product data such as lists of features and pricing information in a playful manner, but it is possible to make it approachable by avoiding the use of jargon or technical terms that assume a level of familiarity or expertise that a new customer may not have. Promotional copy, on the other hand, may place greater emphasis on the energetic, playful, and upbeat aspects of the voice.

A quick note about expectations for voice and tone by content type. All user-targeting aside, there are some common-sense guidelines for evaluating the tone and style of particular types of content. For example:

- Blogs are generally informal, conversational, personal
- Help content should be friendly, encouraging, supportive
- Technical content is generally detailed, serious
- Legal content is, well, legalese

Table 9.1 shows a few examples of what to look for in content written to our example attributes. If your style guide doesn't include a list like this, consider creating one to work with.

Table 9.1 – Examples of what to look for

Attribute	Content characteristic
Friendly	■ Written clearly and conversationally ■ Uses short, simple sentences ■ Uses familiar, common language
Approachable	■ Gives users ways to contact you ■ Content is easy to scan
Conversational	■ Written as if you're speaking to a friend, and want your friend to know what you know ■ Written informally ■ Uses contractions ■ Written in the second person: "you," "your," and "yours"
Energetic	■ Uses the active voice ■ Empowers the customer with action verbs: Find, Search, Explore, Get, Shop, and so on

Editorial/Style Guide

Many organizations have an internal style guide, or follow one of the standards, for issues like use of punctuation, abbreviations, and other terminology. For example, a style guide might address whether to use a.m. and p.m., AM and PM, am and pm, or A.M. and P.M. or when to capitalize or abbreviate certain terms common to the subject matter. When you begin auditing a site, ask your clients whether a style guide exists and familiarize yourself with it so that you can make a single pass through the content and catch all the issues. If the client doesn't have a style guide, consider making or starting one so that you can document the decisions you make as you review content and eliminate inconsistencies.

The purpose of having a style guide, of course, is to ensure a high-quality, consistent reading experience. As content strategists, we understand the importance of quality content. It communicates key messages to the audience and greatly influences visitors' overall experience of the site. It also helps create an impression of the company behind that site.

Allow poorly written, unclear, inconsistent content to stand and you risk leaving the reader with a bad impression of the company. It distracts from the task at hand and can subtly undermine the brand. Insist on making it right and you will not only make a positive impression on the audience, but they will also be more likely to absorb its main messages.

Consistency across your digital communications, as evaluated against brand and style guides, is critical. But many organizations have other, offline touch points with customers as well. Another reason to advocate for consistency across all communication channels is to reinforce the brand credibility and give customers confidence that they are dealing with a single organization dedicated to quality experience.

Content Requirements

Style guides may also address requirements for certain content types to ensure accuracy and consistency. For example, you would typically want product specifications to follow a strict standard and use a consistent set of terminology to describe product features so that accurate comparisons can be made between products.

Site hierarchy also should reflect the way customers think of your products rather than the way the business does. Avoid organizing the site by business unit but instead by what is intuitive to someone outside the company. If your audit process time and resources allow, doing some customer research, such as a card sort to understand how users label and group content, can be very helpful in understanding how users think.

Navigation labels should be evaluated against typical customer tasks and include keywords based on what you know about how your customers search for or think about your site or product. Labels and links should allow a reader to accurately guess at what they lead to.

Plain Language

What if you don't have brand guidelines that include voice and tone or an editorial style guide? You can seldom go wrong by embracing the concept of plain language. According to the Center for Plain Language,

> A document, website or other information is in plain language if the target audience can read it, understand what they read, and confidently act on it.

If the content on your site isn't easy to read, understand, and use, flag it for rewrite.

For a checklist and guidelines for writing in plain language, see the Center for Plain Language[2] website.

Expert vs. Novice

Technical sites or commerce sites that deal with sophisticated equipment have the challenge of speaking both to very expert users and people just getting started.

A new user of a product or technology, for example, may feel unsure about making a decision, lacking confidence in his or her ability to gather the right information. A guided experience, offering ample opportunities to learn more via helpful content, could help enhance their confidence. Look for content that is offered in multiple formats (text, video) and is accessible via multiple entry points.

Sites aren't always easily divided into expert vs. novice content (nor should they be), so be aware that novices can come across technical content. You don't want all the expert-level content written in such a way that it doesn't meet novices' needs, but look for ways that novices can easily navigate to more information and not get stuck at dead ends.

For the expert user, the goal is likely for the content to focus more on making her feel well informed and to offer ways to increase her expertise. If the expert user is a decision maker or if the product appeals to users who already have a high level of knowledge, the user may not need much assistance content but may need comprehensive, specific, accurate information – and may need to find it quickly. Navigation and copy should expressly support this focus by being as brief as possible, and by using precise, concrete, to-the-point language.

---

[2] http://centerforplainlanguage.org

A business audience is more familiar with technical specifications and industry jargon and expects terminology to be used consistently and accurately. Look for content features like specifications, expert forums, and side-by-side comparisons to make it easier for the customer to quickly assess and make decisions.

Summary

There aren't hard-and-fast rules for auditing content against subjective measures like voice and quality. In many cases, the answer to how a particular content set should be authored, presented, and managed is "It depends." Every site is unique, every user is unique, and tastes and standards change frequently. But we still have a responsibility to our readers to make their experience the best it can be, so we work with what we have.

Guidelines, personas, and data provide a framework within which to evaluate content. On top of that framework, we rely on our common sense and hard-won experience as strategists to guide our audit process.

CHAPTER 10
Auditing for Content Effectiveness

Measure what can be measured and monitor what can't, recognizing that
good intentions are no substitute for performance and results.

—Peter Drucker[1]

In This Chapter

In this chapter, we'll look at site analytics data and at how adding it to
your content audit can help you measure content effectiveness.

Measuring Content Success

How do we measure content success? We all want our websites to serve
us well. We want happy, satisfied visitors and we want them to convert,
by whatever our measure – whether it's to become a customer of our
product or service, enjoy our content and share it with others, or engage
with learning content.

To that end, we work to create engaging experiences and high-quality
content; we wrap it in a nice visual package, put it out there, and hope
that it works. But of course, we don't just hope it works – ideally, we
regularly review and measure site effectiveness. Site analytics are widely
available but they aren't usually enough to define effectiveness. For a
more expansive view of content performance, evaluate other measures
as well – direct feedback, customer service and support requests, and
mentions and reviews.

In their paper "Positioning Content for Success: A Metrics-Driven
Strategy,"[2] Kevin Nichols and Rebecca Schneider refer to these two types
of data as "hard" metrics (quantifiable measurements such as conversions, numbers of visits, time on site) and "soft" metrics (qualitative
measures such as user research, behavioral analysis, and customer satisfaction surveys).

To leverage all of this input, a strategy for adding analytics data and
other metrics to your content audit can make the process more manageable, scalable, and insightful.

---

[1] http://www.druckerinstitute.com/project/drucker-institute-forums/
[2] http://9lbikwrikyedoavs8mjf.sapient.com/en-us/sapientnitro/thinking/papers.html

Getting Started with Analytics

Analytics, whether from third-party tools such as Google Analytics or from internal tools for tracking data such as site search metrics, offer us a great deal of data. Key to effectively using analytics to help audit and track your content's effectiveness is deciding which data is most relevant for your needs. In making that determination, think about what your organization values and how that is measured. If sales are a *key performance indicator* (KPI) for your company, tracking conversions is critical. If brand engagement is important, tracking data for social shares and sentiment analysis matters.

Unless you are working on a data-driven inbound marketing strategy or doing search engine optimization, for most content projects, several key categories give us the most actionable input into a content audit – pageviews, exit, and bounce rate. Understanding what these data points mean helps you know what to look for in your analysis.

In brief, the pageviews number is exactly what it sounds like: the number of times a page is viewed. Note that most analytics programs, such as Google Analytics, track both overall pageviews and unique pageviews (the number of visitors to a page). Bounce rate percentage tells you how often the person entered the site from that page and left it directly from that page without viewing any others. An exit percentage, however, shows that the user left the site from that page only after having been elsewhere on the site.

Once you have the data, you may wish to limit your audit. If your site is large and your time or scope limited, you might apply the 80/20 rule and delve deeply into the data only for the top 20% of your visited pages. Or you may decide to focus your effort on your most important pages for conversion – landing pages, product pages, and so on. If you've set up goal pages in your analytics program to track specific user actions such as account creation or purchases, include those in your audit.

Once you've decided how much to audit, add the data to your inventory spreadsheet, and start your review.

Which Data Are Most Meaningful?

If your audit time is limited, it can be helpful to focus your attention on the pages that are receiving the most and the least traffic. With that list in mind, you can review the pageviews data again and find not only how many visits a page has received but the source of the traffic and how long the person stayed on the page once reaching it. Frequency and duration of views is a good measure of loyalty, which is more important for your business in the long run than just visits.

If you have defined conversion paths in your analytics dashboard, test users' browsing path for that conversion activity against an ideal path. What pages are your customers hitting along the way to making a purchase, for example? How long does that pathway take to traverse and how often do they drop off before reaching the call to action? Is it the path you want or expect them to take? If not, you may have identified an opportunity for simplifying your customer journey.

If much of your traffic is driven by organic search, you have a good indication that your content is ranking well for keywords and links – but if the bounce rate is high or the time on page is low, perhaps what they're finding when they get there is not what they were led to expect. High bounce rates may indicate disappointed users and may have negative consequences with your search rank. Use your audit to look for pages with a high number of page views, but also monitor high bounce or exit rates, and review the page content and experience to determine what might be causing that user behavior.

Keep in mind that a high bounce rate may not be a bad thing. Depending on the content of the page, a user may be able to get what she needs in a single viewing and leave satisfied. Looking at bounce rate in conjunction with a metric like time on page or conversion may be more enlightening.

Other traffic sources may include direct links, referrals from other sites, and ad campaigns. This isn't an exhaustive discussion about search engine optimization, so we won't go into the details of how to evaluate different types of traffic; we're more interested in the final result and how to use it to locate which pages to audit.

Does Your Content Have ROT?

Redundant. Outdated. Trivial. If your site has been around for a while, there's a good chance you have some content that fits one or more of these adjectives. Use analytics data to locate pages that have little to no traffic and mark those pages for further review.

While at first blush it may seem that the pages with low traffic are obvious candidates to be weeded out, take this as an opportunity to see whether there is a fixable cause – that is, is the content good but the title misleading? Is it written in an unengaging style or is the information hard to understand? What is the connection to other pages or interactions on the site? If the traffic is low but time on page is high, perhaps the problem isn't with the content, it's because it's hard to find – but worth the effort once reached. In that case you may want to improve your cross-linking strategy by adding additional navigation or links from related content or raise the visibility of the content on the site by positioning it closer to the top level.

Consider Keywords

Most analytics programs tell you what terms users searched on to arrive at your site. Internal search analytics can also offer immediate insight. Try searching your site for those terms – is what you see what you would expect if you were a visitor? If not, how can that content be made more relevant for that term? Perhaps you need to create new content that better matches terms commonly used by your target audiences. Are there any surprises in the list of terms? Perhaps you have identified an opportunity to create content that fulfills what users come to your site to find.

Other Performance Measures

Various measures, such as site analytics, can help you assess content performance, but you should also look at the information you get directly from customers.

Access and review any sources of customer data: direct feedback sent via Contact Us forms or comments on the site; customer service requests, which can be a rich source of information that points out content gaps or inconsistencies; ratings and reviews, which help show what content is popular; and social sharing, which indicates engagement. Search logs are another rich source of information about how customers refer to your products or services (is the terminology they use reflected in your site labeling and content?) and where the gaps in content or navigation lie. If important content is buried so deeply that it has to be searched for to be accessed, you may want to rethink its placement on the site.

A Note of Caution

Analytics data gives you the what, but not the why. Content may be low-performing for many reasons other than its quality. Content placement on the site may not be appropriate for the audience or the stage of the user journey at which it is accessed. The content may be too high-level or too technical for the audience even though it provides valuable information. The solution may be not to remove it but to improve it.

Alternatively, the content may be presented in a format that doesn't work well for the intended audience. For example, burying important support content in unsearchable formats like PDF or video may result in poor traffic or ratings. Again, consider whether another presentation of the content would be more effective.

Summary

Analytics data can be used in conjunction with your other audit measures to create a fully informed view of your content. Look at the highest- and lowest-performing pages as a starting point for identifying where to focus audit efforts. And remember that numbers tell only part of the story.

CHAPTER 11
Auditing Competitor Sites

In This Chapter

In addition to auditing your site against your organization's goals and users, it can be useful to conduct an audit of your competitors' sites to see how yours compares. In this chapter, we look at what to assess on a competitor's site and how to create a scorecard for evaluation.

The Competitive Audit

We all want to create great sites that engage and serve our customers and readers well. So do our competitors. There are many reasons and ways to audit your site content – we've covered personas (Chapter 8), content performance (Chapter 10), and quality (Chapter 9), for example. Auditing it against your competitors' sites gives you an extra lens through which to assess effectiveness and identify gaps and tactical needs.

Even if your organization doesn't have direct competitors, doing a competitive audit can help you assess the larger business or cultural context in which your content lives and provide a way to see best practices in action. For example, if your organization is a nonprofit, you may not have competitors in the sense that an e-commerce site might. But looking at how other nonprofits handle common tasks, like encouraging and processing donations, can help you identify areas for improvement.

Competitive audits can serve another purpose – to convince internal stakeholders of the need for site improvements. Sometimes there is nothing like a side-by-side comparison – evaluating aspects that are important to your organization and showing the gaps and the lower-quality experiences – to make the case.

Getting Started

Much like performing an audit of your own site, the act of auditing another organization's site requires a few basic tasks to start. By now, you've created your own site inventory and audit so you have your baseline for auditing. Using a tool like the Content Analysis Tool, you can run an inventory of the competitive site to use as your audit document. You can also run the tool at Alexa.com on your site and your competitor's to gather data on keywords that send traffic to the site and on other metrics, such as overall popularity.

Before you begin scoring the content, you'll want to do a quick assessment of your competitors' sites. This isn't your actual set of *heuristics* yet, but much like doing your own site inventory and audit, it serves the purpose of familiarizing you with the landscape you'll be assessing. For general-purpose competitive audits, your list could include a range of features to review:

- Audiences
- Type and quantity of content
- Formats
- Language (tone and voice)
- Contributors (numbers, names)
- Community features
- Frequency of publication
- Description/overall impression
- Stand-out or differentiating features

When you've gathered this information, your next step is a side-by-side comparison of your site against your competitors' and an objective evaluation of the comparative strengths, weaknesses, and points of differentiation.

Create Your Scorecard

Begin your audit by determining your goals and the measures (or heuristics) you'll use to score the sites. Put those measures into a scorecard such as the one pictured below. Assess your competitors' websites on the same measures – or on the measures you hope to compete with.

Depending on the site and your goals, the measurements you select may vary. For a blog, you may focus on criteria such as frequency of publication and how often it is commented on, rated, or shared via social media.

For an e-commerce site, you look not only at content depth and breadth – how much information is provided for products? – but also how well other site content is integrated with product content and how well content supports the user journey from evaluation to purchase.

You may want to partner with your user experience counterparts to look at findability, product categorization, and user tasks like product comparison and purchase.

Make sure criteria are specific enough to be measurable, although a certain number of judgment calls are inevitable. To keep the task manageable, you may want to focus on one, two, or three top competitors in your area.

Establish which features or content sets you are assessing, and create your list of heuristics. For each of them, create a set of questions to ask about the site content. An example scorecard for support content might look like Table 11.1:

Table 11.1 – Example of a scorecard

Heuristic	Competitor 1	Competitor 2	Competitor 3	You	Average
Support content is easy to find.					
There are a variety of support options to choose from.					
Support resources are easy to use.					
It is easy to find a physical store location.					

The metrics you include in your scorecard can be numeric (rating on a 1-5 scale, for example) or graphic representations such as Harvey Balls.[1] Numbers make it easier to create averages; visuals may be slightly easier to quickly scan.

Whatever you choose, try to avoid using simple checkmarks or Yes/No since these generally provide too little qualitative information. If the content or your goals don't lend themselves to a literal scorecard, you may also create a written assessment of each site, addressing your audit criteria. This makes side-by-side comparison more difficult but may be a more intuitive way to describe qualitative aspects of content.

---

[1] http://en.wikipedia.org/wiki/Harvey_Balls

Doing the Assessment

Here are some typical heuristics by which to measure content and what
to look for.

- **Breadth and depth:** What is the range of topics covered? For a product
  site, how many products range across how many categories? Are
  articles or product descriptions descriptive and informative?
- **Consistency:** Is content written in a consistent voice, appropriate to
  the audience? Is the writing of consistent quality? Is similar content
  constructed similarly?
- **Completeness:** Do site help functions enable users to get accurate
  answers to questions? Do product descriptions include all the information need to make a purchase decision? Can users access support
  documentation when needed?
- **Currency and Frequency:** Is content up to date? For a blog, how frequently is it published?
- **Findability:** Are the navigational and categorization structures suffi-
  ciently clear that users can proceed with confidence? Is site search
  enabled? Does it include features like synonym matching or support
  for misspellings? Is some content available only behind login? How
  well does search work? Run a few searches for common terms to see
  how relevant the results seem.

Competitive Analysis

You've completed the audit of your competitors' sites and put their
scores or assessments side-by-side with the audit of your site. Now what?

Review the goals you established at the outset of the audit process. If
identifying content gaps was a goal, look at your completeness scores.
For example, do all your competitors include how-to articles to accompany their products? If yours don't, you may want to add that to your
content planning. If your site or products target one demographic, how
well do you compete? How does your content or product range stack
up? Does your writing have a voice appropriate for that audience, or do
your competitors do a better job of hitting that tone? Can people find
content or products on your site? Simple searches and timed browse
exercises can help you determine how discoverable your content is
compared to your competitors.

Conducting these assessments with a clear, objective eye either provides
a set of action items (improve search, simplify navigation, create more
how-to content, and so on) or confirms that you are on the right track
already.

If you do find significant differences or gaps, proceed with caution. It can be tempting to take the findings too literally and jump too quickly to standardize your content and site experience with others' experiences. You still need to maintain your own voice, point of view, and personality. Throw out or improve the bad, but keep the good, and leave space for differentiation.

Summary

For another view on your site's content and experience, compare key scenarios or content experiences with your competitors. The resulting evaluation can be a useful tool in identifying gaps and making the business case for a content improvement project.

CHAPTER 12
Auditing for Global Issues

In This Chapter

If you are preparing for a global content project, such as localizing an existing content set, it's useful to conduct an audit focused on the content issues that may need to be addressed pre-globalization.

The Global Audit

Global sites need to be audited for the same issues as non-global sites – consistency, currency, quality, competitive advantage, and so on – but the global aspect carries with it a new set of ramifications. What makes global content projects different is their potential for harm: a mistake or missed opportunity is problematic enough on a single site but can become exponentially worse when you're talking about multiple sites in multiple languages serving multiple cultures.

A truly internationalized, globalized, and localized website involves the whole team. Developers address issues like Unicode support and separation of code from text. User experience architects think about layouts and text length for UI elements. Designers pay attention to colors and images. Content owners and strategists need to be aware of issues of terminology and cultural relevance.

The *Web Globalization Scorecard,*[1] published annually by Byte Level Research, bases its rankings on these factors:

- **Languages:** Number of languages the site supports
- **Global navigation:** How quickly can users find local content?
- **Global consistency:** Does the site use a global design template across all locales?
- **Localization:** How relevant is the website to the user's culture and country?

According to the scorecard, the top websites in the world support at least 28 languages. They also show a high level of design consistency across their localized sites and use clear global gateways to enable users to select their locale.

Not every business requires a presence across the entire globe and, certainly, many don't have the resources to support localization into 28 or

---

[1] http://bytelevel.com/reportcard2013/

more languages. But even if your site's global ambitions are more modest, the benefits of consistency and relevance are many. Standardization and consistency enable less resource-intensive *localization* – you may be able to use machine translation for some sets of content, freeing you to focus customization efforts on the content that is most valuable for cultural relevance.

Optimizing for Localization

Writing content to be localized forces some general business writing best practices, including following standard sentence structures; using clear, non-ambiguous terms; writing in the active voice; following standard rules of punctuation; spelling out acronyms; and avoiding colloquialisms and jargon.

Design is another important aspect of localization. Special attention needs to be paid to the use of colors, icons, and images because of the potential for misunderstanding or offensiveness.

Consistency

If a site has not yet been localized or globalized, there is an opportunity to catch issues early and prevent them from later becoming more costly and difficult to manage. For example, consistency increases user comprehension and confidence. When looked at in the context of a global site, however, consistency has an additional financial implication – the less variation in terminology, the fewer terms that need to be localized. Take the term for the concept of proceeding to the next step in a process – Enter, Submit, and Go being common instances. These terms (often on buttons in the interface) are often used interchangeably and inconsistently. But standardizing on a single term saves localization time, money, and design time.

Consistency also lessens the opportunity for errors in translation (whether human or machine-driven), which affect user perception and can even misrepresent important information.

Content Length

According to localization vendor Lionbridge, "On average, an English sentence, when translated, is 30–40% larger in some European languages (and can be as much as 200% for a particular word). With Asian languages, text may shrink by 20–50%." (See the presentation *Building a Global Web Strategy*[2] for more detail.)

---

[2] http://www.slideshare.net/Lionbridge/building-a-global-web-strategy

Using fewer, simpler terms that can be consistently translated across languages saves time and money and reduces the potential for mistranslation. Look for opportunities to tighten your copy.

Content length is a major consideration for user interface elements as well. Consider how text expansion or contraction affects buttons, navigation labels, tabs, and tabular content.

Cultural References, Colloquialisms, and Jargon

There are oft-cited instances of English expressions or terms, often used in product names or slogans, translating poorly. The Chevrolet Nova car, for example, purportedly did not sell well in Mexico because in Spanish, Nova sounds like "no va," which means, "Doesn't go."

Machine translations are particularly vulnerable. Linguists use the term conceptual equivalence to describe translation that ensures that exact meaning is retained. As the content auditor, you may not be able to influence brand names or slogans, but you can be aware that brand names and slogans may require special handling, and you can flag them for human localization.

Idiomatic expressions create problems for translation, too. For example, common American office expressions such as "run it up the flagpole" or "around the water cooler" may not be meaningful in all cultures.

Similarly, avoid references to holidays or seasonal weather. For example, winter in Brazil is different from winter in Scandinavia. References to snow and Christmas are irrelevant in Brazil and insensitive in non-Christian cultures.

Terminology

Localization goes beyond translating words. If your organization has an editorial style guide, does it include entries for content elements that are likely to arise in a global content set? For example, the following must all be addressed, with guidance for consistent use and terminology that works across cultures.

- Currencies
- Date and time
- Measurements
- Telephone numbers
- Sort order
- Capitalization
- Postal codes

Language and Communication Style

In their book *The Culturally Customized Website: Customizing Websites for the Global Marketplace*, Nitish Singh and Arun Pereira write of the differences in perception and language across cultures. They divide cultures into low context (example: the United States) and high context (example: Japan). High-context cultures, featuring close connections among group members, make greater use of symbols and nonverbal cues to communicate. Low-context cultures are "logical, linear, and action-oriented," and information is explicit and formalized. In a low-context culture, therefore, a hard-sell approach is acceptable, including the use of superlatives, emphasis on product advantages, and direct comparison with competitors. In a high-context culture, language is generally more polite, indirect, and humble.

Awareness of cultural differences such as these, and many other aspects of Singh and Pereira's research, provides another lens through which to assess your global site content.

Design

From a design perspective, review policies for use of the following:

- Colors
- Icons
- Images

Colors carry different significance in different cultures. In America, red indicates stop, danger, or importance; in Greece, love and good luck; in Africa, death and bloodshed. For more information on the cultural significance of colors, see the book *Beyond Borders*, by John Yunker.

Icons, being pictorial and therefore not requiring localization, can be immensely useful in conveying information – as long as they aren't culture-specific. An American-style mailbox, for example, may not work as an icon in cultures without a tradition of postal mail delivery.

Photographs, particularly those containing images of people, should be looked at carefully. Cultural differences in dress and gestures can make a seemingly innocuous image offensive. For example, look at the image in Figure 12.1, from laundry machine manufacturer Speed Queen.

Figure 12.1 – Speed Queen website

Notice that the woman in the photo is wearing a sleeveless shirt. In Western cultures, this image would be unremarkable, but in other cultures it may be offensive or at least culturally insensitive. If providing localized versions isn't feasible, consider using images that are either more conservative or that don't include people at all.

Another best practice: avoid embedding text in images and videos to limit translation needs, a good idea in general, since text that is part of an image is not indexable by search engines and is not easily edited.

Search Engine Optimization (SEO)

Keywords come into play for localization efforts too. Each localized site should also have its own set of localized keywords. Review content metadata for the presence of keywords that are linguistically correct translations and that include local variants and synonyms.

Video and audio transcripts should also be localized to improve search and to provide a more accessible experience for the reader.

The Global Challenge

The challenge is to retain life and color in your language and have your content be culturally relevant without creating problems for your localizers or confusion for your users. If regional resources are available to create and maintain sections of the site that lend themselves to local content and style, decentralizing management of those sections is ideal. If centralized management is the reality in your organization, you may need to write around those references or find culturally neutral ways to express the same concepts.

Summary

Global websites offer special challenges for the content auditor. In addition to the typical issues we look for in any audit, we need to take a larger view of how content is managed and presented in multiple cultures and languages. Identifying consistency, terminology, and design issues early can save time and trouble in the future when your site is localized.

CHAPTER 13
Auditing for Legal or Regulatory Issues

In This Chapter

An often overlooked aspect of content auditing is looking for the kinds of issues that can cause legal problems for an organization. In this chapter, we look at some of the content-related issues that a typical content audit might include.

Conducting an Audit for Legal Issues

What would a lawsuit or a period of disruption cost your business – not just in financial terms, but in terms of public relations and brand identity? In addition to other ways we audit websites, here are some examples of the kinds of issues that are important to review because of their potential to harm the business.

A legal audit of a website involves a long checklist of issues outside the purview of a content strategist or site manager – see "Benefits of a Legal Audit (with checklist)"[1] for an example – but some content-related issues we must be able to spot. As with any other form of site auditing, partner with the appropriate personnel in the organization – in this case, the legal department – and gather and review all existing policies and procedures before beginning.

Note: This chapter is not intended to replace the advice of your legal team nor provide guidance for a comprehensive legal audit. The following information is provided as a guide to some of the issues that a content strategist or site manager might review and bring to the attention of legal counsel.

Copyright

Any site content that your organization did not generate – including text, images, audio, video, and documents – must have rights secured to avoid copyright infringement. *Digital asset management systems* (DAMs) usually include rights management features, which can be reviewed periodically to ensure that no rights have expired. Work with your content and creative teams to understand how assets for the site are sourced and managed. If your company does not have written policies for use of third-party material, it may be helpful to all to create them.

---

[1] http://www.exec-counsel.com/content/benefits-legal-audit

If your site has a copyright notice in the footer, check the date to make sure it's current. You can also register your site with the United States Copyright Office[2] to protect your own material.

Trademark

In general web content, trademarks are less commonly seen these days, but your organization may require them. If so, your legal department can provide you with a list of your company's trademarks (™), service marks (℠), and registered trademarks (®), and the guidelines for their use. As you review content, look for correct usage of these marks.

Privacy Policy and Terms of Use

In a world of rapidly changing expectations about access to and use of personal information, it is critical to publish your policies related to what types of user information is collected, how it is used, how it is secured, and whether and how users can update or remove their information.

Privacy policy and terms of use are often of the "set it and forget it" type of content – once they're posted, and a small-point-type link placed in the footer, they can languish unreviewed for years. Ask your legal counsel to review your policies, particularly after the implementation of any new site features that involve creating user accounts, purchasing or donating, or gathering user-generated content.

If you do e-mail campaigns or newsletters, make sure that users must opt in and that they can easily opt out.

User-Generated Content

User-generated content presents a special set of risks. In the absence of strict moderation, users may post offensive content, links to inappropriate sites or malware sites, copyrighted material, and much more. If your site features user-generated content – reviews, comments on blogs or articles, user photo galleries, and the like – you must have a clear, easily found policy stating the user's rights and restrictions and your company's rights with regard to blocking or removing content, using material for commercial purposes, and so on.

It is common to require users to accept the terms of the policy before posting; if that is not now the case on your site, it might be a good idea to put that into place. And if you do have a policy, make sure it is current with site features and that your legal counsel regularly reviews it.

---

[2] http://copyright.gov

Regulatory Issues for Nonprofits

Nonprofit organizations have special requirements for public disclosure and retention of materials such as board meeting minutes, annual reports, and financial reports, and for the use of their websites for distribution of these. If you work on a nonprofit site, check with your legal counsel as to how long these materials must be available.

If the site includes fundraising, there are additional best practices and requirements. See the tips at the Nonprofit Law Blog.[3]

Links

An often overlooked area of risk is linking. Does your company have a policy for links out to ensure that your site doesn't link to inappropriate sites or to sites that have copyright-infringing material? A simple fix is to make it clear when a link is to an external site, and include a notice in your legal policies that disclaims responsibility for the content of any sites linked to from yours. As with all of the above, consult your legal counsel to see if this is an area of concern and, if so, what to look for.

Summary

A comprehensive audit must include a pass through the content for copyrights, trademarks, privacy policies, and other legal or regulatory issues that might open the organization to scrutiny or cause harm. The content auditor should partner with the organization's legal counsel to ensure that all appropriate issues are addressed.

---

[3] http://www.nonprofitlawblog.com/home/2010/09/top-5-fundraising-legal-tips.html

CHAPTER 14
Presenting Audit Findings

In This Chapter

Just as important as gathering your audit data and creating recommendations and strategies is presenting that information in a way that drives action. In this chapter, we discuss what to include in your findings and how to present those findings for maximum effectiveness.

Developing a Strategy for Your Strategy

An audit is the result of gathering data, processing information, and arriving at conclusions. By the time you've completed your audit, you have likely become the most knowledgeable person on the team about the site content. You know what's there, how it's organized and structured, how well it's written, how your customers or readers are using it, and more. You also have a strategy for what needs to happen next – a plan for revision, a list of content to be created, and a migration plan.

If you are the business owner of the site, you can proceed directly to implementation. But if you are presenting to another audience, you need to find effective ways to communicate your findings and strategy.

One of the major challenges in presenting the results of a content audit is in extracting and communicating a message from the audit data. The larger challenge, though, is doing so in a compelling enough manner that the audience pays attention and your goals are achieved.

In essence, you need a content strategy for your content strategy.

- Identify your audience, the context, and your message.
- Choose a delivery mechanism that works for your audience.
- Present your findings in a timely way for project decision making.

Telling the Audit Story

A content audit presentation should tell the story of what you found, both in summary and in detail. The audit likely has multiple audiences within the organization, so pick your form factor and level of detail accordingly.

As discussed previously, if you present the case for kicking off a redesign or content strategy project to executive sponsors, it may suffice to summarize your findings, explain the scope of the proposed project, and list general recommendations. If you are presenting to a broader team – for example the business owners of the content, the user experience and content teams, and the technical team – then your presentation requires more detail.

The audit should be written in sufficient detail that a reader who did not conduct the audit understands the context and findings. Tell your audit story orally (to spark conversation, answer questions, and identify issues that need follow-up) and in writing (to provide detail for your audience to dive into on their own).

Presenting the Findings

Knowing the goal of your presentation – the actions you want people to take based on your findings – is critical to deciding how to share the information. Are you asking for resources to initiate a project? Or are you using the audit as the starting point for a dialogue with the team?

Who is your audience and how much background do they have about the project? If your audit presentation aims simply to introduce a project, you may spend more time on the methodology and high-level findings, sending a detailed document after the presentation. If the purpose of the audit presentation is to leave the room with decisions made and resources allocated, distribute the document first, and ask that all attendees read it and come prepared to discuss it in detail.

Your audit presentation needs a clear call to action. It is not enough to simply catalogue the issues, make recommendations, or discuss benefits. What is the desired outcome, and how does that involve or affect your audience? Everyone should leave the room (or put down the document) knowing what is being asked of them in terms of both time and resources. Everyone needs to understand the timeline for implementing a plan, the issues, the risks, and the next steps.

Just Say No to Sharing Spreadsheets

Spreadsheets are great for collecting data but they don't lend themselves to presentations. They can overwhelm and obfuscate the message. Looking at row after row and column after column of data, it's hard to determine what's most important. What are the key takeaways? Where should the implementation plan be focused? You can color code and format the cells, but that's no substitute for an easy-to-scan document or graphic. And even the most creatively formatted spreadsheet is usually still just a collection of information, not an analysis.

For that reason, I suggest creating a written audit, including illustrations and examples – perhaps some infographics for visual punch – and a slide presentation.

Tosca Fasso's two-part series "Audits to Insights" on the Content Insight blog discusses creating data visualizations as an effective way of presenting audits, with helpful advice for the non-designer. In Part 1, "Keep Your Spreadsheet to Yourself,"[1] she says,

> As much as words are a safe place for many of us, research has shown that people process visuals 60,000 times faster than text, and in a busy work environment, that makes a difference. The fact that images are processed so differently has even been given its own name: PSE – Pictorial Superiority Effect. Tests have shown that people can recall – with 90% accuracy – more than 2,500 pictures several days later and with 63% accuracy a year later. So if you're trying to make an impression, words alone will fall short.

Focus on Persuasion

A thorough content audit results in a great deal of information, not all of it relevant and not all of it persuasive. When presenting findings with the goal of driving change, select the data points that your audience will find most convincing. Do your recommendations save the company money? Drive more sales? Create internal efficiencies? Choose information that illustrates those points, presenting it in a compelling way – via numbers, charts, and diagrams if appropriate – to achieve your goals.

Creating the Audit Document

Overview

A content audit document generally begins with an overview that addresses the objective (what was assessed and why), the methodology (the process and criteria used), the general findings, a section-by-section assessment, a summary of recommendations, and a proposed plan.

Related: Appendix G, *Content Audit Template*

Assessment

Most websites comprise sections that have different owners, audiences, purposes, and styles. You may even have used different criteria to audit each section. Your audit document should include your observations and recommendations for each area and each audience. Depending on

---

[1] http://www.content-insight.com/blog/2013/08/audits-insights-part-1/

the perspective you took as you conducted your audit, this section of your document may be organized by audience, section, or content type.

Unless you're auditing a single-purpose site, such as a blog, there are multiple content types – landing pages, product pages, press releases, and so on. Regardless of the site section they appear in or who owns them, these can be discussed in aggregate and lend themselves to an "apples to apples" comparison.

Recommendations

Finally, every audit needs a summary of recommendations and a proposed plan. Make the recommendations measurable, achievable, and time-based, keeping in mind the organization's resources, technology capabilities, and the context for the audit. Show pros and cons of each recommendation, backing them up with numbers. Give projected revenue increases or efficiency improvement numbers to bolster your case.

Next Steps

Provide budget estimates for a follow-up project, including projected resources and tools investment. Propose project milestones and a timeline. List the decisions that need to be made, by whom, and by when.

Summary

Presenting your findings in a compelling way, backed up by data, and a well-thought-out set of recommendations is key to proving the value of your content audit project and getting buy-in for follow-on improvement projects. Select data carefully, and present the information in an easily consumed manner, keeping in mind your audience and the decisions that you are asking them to make.

Audit Presentation Essentials

- Summarize the current state
- Describe the future state
- Analyze the gap between current and desired states
- Discuss risks and opportunities
- Provide next steps and recommendations

CHAPTER 15
The Ongoing Audit Process

In this Chapter

The content inventory and audit play an important role in the ongoing governance of a website's content governance process. Instituting the practice of ongoing inventories and audits can help content teams ensure a consistent level of quality. This chapter presents a brief overview of what governance is and how to use inventories and audits to support governance initiatives.

The Role of Content Inventory and Audit in Governance

Website governance covers a broad range of policies, standards, and structures for creating and maintaining data, content, and applications. In this book, I don't cover all the complexities of site governance (for resources on governance, see Additional Reading), but I would like to briefly address content governance and some ways that an inventory and audit can play a part.

Content governance is often expressed as lifecycle management – the rules and processes that underpin everything from content planning to creation to publication to ongoing optimization. The roles and tasks that accompany those steps include identifying who is responsible for creating and maintaining content, developing standards for content quality, and incorporating metrics and feedback into a process of continual improvement.

When governance policies are not in place or are not followed, website content can become disorganized, stale, and ineffective at meeting business and user needs. These problems can trigger a content strategy initiative when the business realizes that the site is failing. A time-consuming, expensive project gets kicked off, an inventory and audit are completed, and a strategy is developed. To avoid costly one-time improvement efforts like this, you need to create a "virtuous circle" – a feedback loop that enables your company to learn and improve over time. You need to update your style guides, your glossary, and your governance policies, and then feed all that back to your content creators so that new content is created to updated standards and you're constantly improving rather than doing major overhauls.

The Rolling Inventory and Audit

How do you create that virtuous circle? Institute a rolling (ongoing, periodic) inventory and audit. A rolling inventory and audit allows you to assess content mix, quality, and effectiveness against ever-changing audience needs and business goals.

The inventory, done at regular intervals or after major content publishing initiatives, enables you to monitor the quantity and types of content on your site. As discussed in Chapter 4, the data you gather in an inventory, particularly if you are using an automated tool, can help you quickly identify trouble spots, such as missing metadata, unwieldy site structure, and problematic metrics. The inventory also gives you the structure to track information, such as the content owner and the age of the content, that helps when you audit.

Content planning, often the first step in a content lifecycle, can benefit from the inventory too, as you track what content exists, what's effective, and what's not, helping you plan to fill gaps or strengthen weak areas.

At the other end of the cycle, the data supports ongoing optimization of content as you analyze your metrics to see what should be pruned or revised.

The content audit can also be done on an ongoing basis. You probably don't have the resources – nor is there a need – to audit every piece of content frequently. Instead, identify the content areas that are most likely to stray from your quality standards, either by becoming outdated or by no longer adhering to your brand guidelines.

For example, seasonal content must be reviewed at the end of each season. But there is no need to regularly revisit published press releases other than to consider archiving them after a certain number of months or years. Content that changes frequently should also be reviewed frequently – for example, content about products and services. Content that tends to be overlooked because it is considered static or not directly related to sales or other conversion metrics, such as company information and staff pages, should also be reviewed regularly.

Keeping track of your content's age and setting a reminder to review any content older than, for example, a year is one way to trigger an audit exercise. You can also plan audits around your editorial calendar.

A rolling audit is also a great way to draw upon the larger content team. Just as you assembled a team to do the initial audit, dividing up responsibilities, you can assign team members ongoing audit duties, breaking up the audit by content area, for example. This not only distributes the workload but also helps ensure ongoing involvement with content quality and buy-in to the process across the organization.

Summary

Websites are living entities, constantly changing and adapting to new business strategies and new audiences. Organizational energy is often focused more on the creation of new content than on the governance and ongoing maintenance of existing content. The result can be sites that are overgrown and no longer effective at meeting goals. Rather than let your site get to the point where a major content repair project is required, adopt the rolling inventory and audit to keep the site in a state of constant review and improvement.

CHAPTER 16
Conclusion

Inventory, Audit, Analysis

Conducting content inventories and audits are valuable exercises for understanding the quantity and quality of your content. But it's the analysis of the outputs from those processes that supports creating effective and sustainable content strategies. Just as we looked at the business process improvement framework DMAIC for an organizing structure for this work, we can look at the Reflective Learning Cycle as a model for ensuring that we turn what we learn into action. The model, developed by professional development coach John Driscoll,[1] asks three questions: What? So what? and Now what?

- The content inventory is the *what*: What do we have? What is our starting point?
- The audit is the *so what*: What is wrong with it? What are the implications?
- The analysis is the *now what*: What do we do with what we've learned? What actions do we take? What change do we effect?

Select Your Tactics, Know Your Goals

This book has presented a number of tactics and techniques for planning, scoping, and conducting content inventories and audits. Effective methodologies are important, but they alone can't achieve the larger goal: creating and maintaining high-quality content.

What is high-quality content? It is content that fulfills business goals and user needs and inspires ongoing audience engagement and loyalty. When we set out to evaluate and improve our content, we need to be grounded in an understanding of those goals and needs. From the business side, we need to understand the context in which our project exists, including the metrics by which success is measured and the readiness of the organization to make and sustain change. As representatives of the content's users, we need to understand who they are, what they care about, why they are interacting with the content, and what tasks they need to complete.

The content inventory and audit are means to an end: the development of a content strategy and user experience that delivers for both the business and the users. We begin content projects, whether a content

---

[1] http://www.supervisionandcoaching.com/

refresh or a website migration, by learning everything we can about where we are. To map a path to Point B, we need to locate Point A. Understanding where your content lives, how much of it you have, how it is organized, how it is written, and how your users engage with it sets the stage for an informed strategy. Using both quantitative and qualitative data analyzed against defined standards and extracted actionable insights creates a strong case that can be communicated to decision makers.

Learn By Doing

This book tells you how to plan for, scope, and conduct content inventories and audits. Many of the concepts and processes discussed have been around for some time, so there is more information available if you choose to do more reading. Some of the best resources are listed in Additional Reading. But, to quote a well-known sporting goods company, at some point you need to just do it. Jump in, learn, refine your process. Do it again. Your circumstances dictate how you conduct your content analysis activities, so adapt and modify what you've learned here to your situation. Then share your experiences with the community so that we can all continue to learn and develop our skills.

Inventory and Audit Resources

A selected set of example deliverables related to the content inventory and audit and recommended reading for further exploration.

APPENDIX A

Content Inventory Spreadsheet

Content Insight Site Inventory

Location	Type	Title	Description	Keywords	H1TagTexts	WordCount	PageViews	ExitRate	Bounces	Size	LinksIn	LinksOut	Images	Audios	Videos	Documents
https:/ / example.com/	text/ htm	Content Insight	Content Inventory and Analysis Made Easier	Content Insight provic	Turn inventor	329	1164	33.4192	266	13036	270	49	2	0	0	0
http:/ / example.com/ about- us.html	text/ htm	About Us | Content Insight	Content Insight, home of the Content Analys	Content Insight, Paula	About Us	556	42	28.5714	1	14434	260	41	2	0	0	0
http:/ / example.com/ blog.html	text/ htm	Blog | Content Insight	Home of the Content Insight blog, featuring	content inventory, co	Blog	1966	143	23.7762	8	54618	272	79	2	0	0	0
http:/ / example.com/ cat- pricing.html	text/ htm	Content Analysis Tool Pricing | C	Select a subscription plans and start creatin and easily.	Content Analysis Tool	Pricing Plans	573	436	34.1743	38	14941	264	40	2	0	0	0
http:/ / example.com/ privacy- policy.html	text/ htm	Privacy Policy | Content Insight	Privacy policy for the Content Insight LLC we	Content insight	Privacy Policy	1978	11	45.4545	0	20869	260	36	2	0	0	0
http:/ / example.com/ products.html	text/ htm	Content Analysis Tool | Content I	The Content Analysis Tool (CAT) quickly and	Content Analysis Tool	Content Analy	729	1306	45.4058	405	16988	272	48	2	0	0	0
http:/ / example.com/ resources.html	text/ htm	Content Inventory and Audit Reso	Learn about the Content Analysis Tool (CAT)	Content Analysis Tool	Resources	828	255	18.0392	8	16390	272	67	2	0	0	0
http:/ / example.com/ report- bug/	text/ htm	Report a Bug | Content Insight	Contact Content Insight to report issues wit	Content Analysis Tool	Report a Bug	184	4	31	8	0	0	0	1	0	0	0
http:/ / example.com/ request- demo/	text/ htm	Request a Demo of the Content A	Request a personalized demonstration of th	Content Analysis Tool	Request a Der	226	19	33	8	0	0	0	1	0	0	0
http:/ / example.com/ request- feature/	text/ htm	Request a Content Analysis Tool	Contact Content Insight to share your featur	Content Analysis Tool	Request a Fea	202	3	24	9	0	0	0	1	0	0	0
http:/ / example.com/ resources/ cat- video- tu	text/ htm	Video Tutorial for the Content An	Watch a short video about the features and	Content Analysis Tool	CAT Video Tu	668	173	37	14	0	0	0	1	0	0	0
http:/ / example.com/ resources/ tour- cat/ das	text/ htm	Content Analysis Tool Dashboard	Annotated screenshot of the Dashboard in t	Content Analysis Tool	Dashboard	196	5	36	9	0	0	0	1	0	0	0
http:/ / example.com/ resources/ tour- cat/ job	text/ htm	Content Analysis Tool Job Details	Annotated screenshot of the Job Details view	Content Analysis Tool	Job Summary	197	5	36	9	0	0	0	1	0	0	0
http:/ / example.com/ resources/ tour- cat/ job	text/ htm	Content Analysis Tool Job Setup |	Annotated screenshot of the Job Setup form	Content Analysis Tool	Job Setup	179	4	35	9	0	0	0	1	0	0	0
http:/ / example.com/ resources/ tour- cat/ res	text/ htm	Content Analysis Tool Resource D	Annotated screenshot of the Resource Detail	Content Analysis Tool	Resource Deta	226	4	34	9	0	0	0	1	0	0	0
http:/ / example.com/ terms- use/	text/ htm	Content Analysis Tool Terms of U	Content Insight provides tools for Content S	Content Insight	Terms of Use	4425	36	33	7	0	0	0	0	0	0	0
http:/ / example.com/ resources/ articles/ cont	text/ htm	What is a Content Inventory? | Co	Introduction to the content inventory, a qua	content inventory, Co	What is a Con	765	93	51.6129	28	13857	3	42	2	0	0	0
http:/ / example.com/ resources/ articles/ crea	text/ htm	Creating Ef ective Content Invento	Knowing your goal, audience, and timeline c	content inventory, co	Creating Ef ec	1084	17	11.7647	1	15298	5	43	2	0	0	0
http:/ / example.com/ resources/ user- guide.h	text/ htm	User Guide to CAT | Content Insig	Guide to using the Content Analysis Tool (C	Content Analysis Tool	User Guide to	3114	55	41.8182	8	32826	32	45	2	0	0	1
http:/ / example.com/ resources/ articles/ web	text/ htm	Website Content Audits | Content	By transforming a content inventory into a c	content audit, content	Website Conte	1436	311	64.3087	174	20966	5	59	2	0	0	0
http:/ / example.com/ resources/ articles/ web	text/ htm	Website Content Inventories | Co	The key to successful website content maint	content inventory, Co	Website Conte	1252	375	60	192	18748	12	53	2	0	0	0
http:/ / example.com/ resources/ articles/ web	text/ htm	Website Content Tracking | Conte	In maintaining and managing a site, keeping	content inventory, co	Website Conte	1197	85	58.8235	42	18032	5	49	2	0	0	0
http:/ / example.com/ resources/ articles/ cat-	text/ htm	CAT for Site Managers | Content I	The Content Analysis Tool (CAT), provides a	website management,	CAT for Site M	633	5	0	0	12930	2	43	2	0	0	0
http:/ / example.com/ resources/ articles/ cat-	text/ htm	CAT for Information Architects | (	For an information architect, a content inve	content inventory, inf	CAT for Inforr	562	8	0	0	12642	2	44	2	0	0	0
http:/ / example.com/ resources/ articles/ cat-	text/ htm	CAT for Content Strategists | Con	The content inventories created by the Cont	content inventory, Co	CAT for Conte	569	11	0	0	12891	2	47	2	0	0	0
http:/ / example.com/ resources/ articles/ job-	text/ htm	Power Tips: Job Setup | Content I	Set up your site crawl in the Content Analysi	content inventory, Co	Power Tips: Jo	1225	14	42.8571	2	18343	3	46	2	0	0	0
http:/ / example.com/ resources/ articles/ tour	text/ htm	Take the CAT Tour | Content Insi	Take an annotated walkthrough of the features of the Content Ana		Take the CAT	339	148	3.37838	1	12108	42	47	2	0	0	0
http:/ / example.com/ resources/ 42- power- ti	text/ htm	Power Tips: The Dashboard | Content Insight			Power Tips: T	843	12	16.6667	1	14692	2	42	2	0	0	0
http:/ / example.com/ resources/ 43- why- cat.	text/ htm	Why Use CAT? | Content Insight			Why Use CAT?	710	24	25	2	13403	7	45	2	0	0	0
http:/ / example.com/ resources/ 44- power- ti	text/ htm	Power Tips: CAT Exports | Content Insight			Power Tips: C	650	6	0	0	13375	2	42	2	0	0	0
http:/ / example.com/ resources/ 46- template	text/ htm	A Template for Content Inventory and Audit | Content Insight			A Template fo	1770	34	32.3529	0	20328	3	47	2	0	0	0
http:/ / example.com/ resources/ 48- using- pe	text/ htm	Using Personas in Content Audits | Content Insight			Using Persona	1718	47	51.0638	17	20963	9	50	2	0	0	0
http:/ / example.com/ resources/ 49- the- qual	text/ htm	The Qualitative Content Audit | Content Insight			The Qualitativ	2304	133	54.8872	54	24379	7	46	2	0	0	0
http:/ / example.com/ resources/ 50- the- com	text/ htm	The Competitive Content Audit | Content Insight			The Competiti	1321	64	62.5	34	18155	4	48	2	0	0	0
http:/ / example.com/ resources/ 51- who- doe	text/ htm	Who Does Content Audits? | Content Insight			Who Does Co	1137	6	0	0	16145	2	46	2	0	0	0
http:/ / example.com/ resources/ 52- adding- a	text/ htm	Power Tips: Adding Analytics to Job Setup			Power Tips	544	21	42.8571	4	13858	7	43	2	0	0	0
http:/ / example.com/ resources/ 53- power- ti	text/ htm	Power Tips: Views and Custom Columns			Power Tips	608	18	5.55555	1	13300	7	42	2	0	0	0
http:/ / example.com/ resources/ 62- the- roi- (	text/ htm	The ROI of Content Inventories and Audits			The ROI	1433	42	57.1429	22	17708	6	47	2	0	0	0
http:/ / example.com/ newsletter/ 58- content-	text/ htm	Content Insight Newsletter - July 2013			Content Insig	914	4	25	1	31032	1	57	2	0	0	0
http:/ / example.com/ newsletter/ 60- content-	text/ htm	Content Insight Newsletter - August 2013			Content Insig	878	3	0	0	29177	1	55	2	0	0	0
http:/ / example.com/ newsletter/ 61- content-	text/ htm	Content Insight Newsletter - September 2013			Content Insig	708	5	20	1	30033	1	56	2	0	0	0
http:/ / example.com/ newsletter/ 63- content-	text/ htm	Content Insight Newsletter - October 2013			Content Insig	828	4	25	1	26828	1	52	2	0	0	0
http:/ / example.com/ newsletter/ 64- content-	tB48ext/	Content Insight Newsletter - November 2013			Content Insig	781	5	40	2	28230	1	57	2	0	0	0
http:/ / example.com/ newsletter/ 65- content-	text/ htm	Content Insight Newsletter - December 2013			Content Insig	721	8	25	0	25091	1	48	2	0	0	0
http:/ / example.com/ report- a- bug.html?view	text/ htm	Report a Bug			Report a	688	10	30	2	33040	3	40	2	0	0	0
http:/ / example.com/ support- cat.html	text/ htm	Support CAT			Support CAT	326	3	33.3333	0	10801	4	39	2	0	0	0
http:/ / example.com/ terms- of- use.html	text/ htm	Terms of Use			Terms of	4461	30	40	4	36405	260	40	2	0	0	0
https:/ / example.com/ contact.html?view=for	text/ htm	Contact Us			Contact Us	712	22	13.6364	1	32071	8	45	2	0	0	0
https:/ / example.com/ log- in.html	text/ htm	Log In			Log In	198	483	7.66046	30	8836	528	36	2	0	0	0
https:/ / example.com/ newsletter.html	text/ htm	*The Content Insight Newsletter*			The Content	254	20	0	0	12344	268	46	2	0	0	0
https:/ / example.com/ newsletter/ 57- conten	text/ htm	Content Insight Newsletter - June 2013			Content Insig	699	4	0	0	29731	1	58	2	0	0	0
https:/ / example.com/ resources/ 45- turning-	text/ htm	From Inventory to Content Matrix			From Inventor	864	18	33.3333	4	14444	1	47	2	0	0	0
http:/ / example.com/ blog/ 2013/ 03/ 06/ 21-	text/ htm	Content Insight News			Content Insig	1673	12	58.3333	5	25766	8	56	2	0	0	0
http:/ / example.com/ blog/ 2013/ 07/ 24/ 42-	text/ htm	Content Insight News			Content Insig	795	3	0	0	20636	3	64	2	0	0	0
http:/ / example.com/ blog/ 2013/ 08/ 05/ 43-	text/ htm	Content Insight News			Content Insig	669	3	0	0	17971	5	50	2	0	0	0
http:/ / example.com/ blog/ 2013/ 08/ 13/ 45-	text/ htm	Content Insight News			Content Insig	815	3	0	0	18989	5	51	2	0	0	0
http:/ / example.com/ blog/ 2013/ 08/ 20/ 46-	text/ htm	Content Insight News			Content Insig	1293	4	0	0	22772	12	51	2	0	0	0
http:/ / example.com/ blog/ 2013/ 08/ 26/ 47-	text/ htm	Content Insight News			Content Insig	1070	4	0	0	22390	5	54	2	0	0	0
http:/ / example.com/ blog/ 2013/ 09/ 30/ 51-	text/ htm	Content Insight News			Content Insig	597	5	0	0	17882	1	52	2	0	0	0
http:/ / example.com/ blog/ 2013/ 10/ 16/ 52-	text/ htm	Content Insight News			Content Insig	1181	4	0	0	21697	4	51	2	0	0	0
http:/ / example.com/ blog/ 2013/ 11/ 11/ 53-	text/ htm	Content Insight News			Content Insig	887	4	0	0	21124	1	49	2	0	0	0
http:/ / example.com/ blog/ 2013/ 11/ 19/ 54-	text/ htm	Content Insight News			Content Insig	1083	11	72.7273	6	21184	1	50	2	0	0	0
http:/ / example.com/ blog/ 2013/ 12/ 12/ 55-	text/ htm	Content Insight News			Content Insig	407	11	45.4545	4	17055	1	60	2	0	0	0
http:/ / example.com/ blog/ 2013/ 12/ 18/ 56-	text/ htm	Content Insight News			Content Insig	629	14	50	3	19116	4	60	2	0	0	0
http:/ / example.com/ blog/ 2013/ 12/ 27/ 57-	text/ htm	Content Insight News			Content Insig	895	8	37.5	1	20187	1	59	2	0	0	0

APPENDIX B
Stakeholder Interview Template

Questions for Business Owners

Role

- What is your role in the organization?
- What are your major responsibilities?
- What critical issues face the organization?

Goals

- What is the primary purpose of the website from an organizational perspective?
- What are your objectives for the site for the coming year?
- Where is the real opportunity? Why?
- What is the role of the site in reaching your audiences? How does it work with other content channels?
- How do you judge website success?

Users

- Who should the site target? Who should not be targeted?
- Who are your primary stakeholders? How do their needs differ from those of your target audience?
- What is the primary purpose of the site from your users' perspective?
- What content and features do users expect to find on the site?
- What concerns might users have about the organization, and how does the site address them?
- What user needs and expectations are now being met? What user needs and expectations are not now being met?
- What do you want users to think and feel after visiting the site?
- Have you conducted research that would be relevant to users' expectations of the site? If so, can we review?
- Where is the user being best served on the current site? Underserved?

Best Practices

- Who are your peers (as in similar businesses/organizations)?
- Which peers have a good website? What do you like about their sites?
- What trends in your industry worry or inspire you?

Editorial Issues

- From an editorial perspective, what content and features do you see as providing the greatest value to your readers? The least?
- Do you have a style guide?
- How important is social media for your organization? How do you leverage social media on the site?

Questions for Content Creators and Site Management

Site Content Maintenance and Development

- How frequently is the site updated?
- What are the pain points with content publication?
- Who creates the content?
- Which legal and regulatory requirements dictate the need for particular types of content or content retention?
- What is your typical workflow for identifying a content need, getting it created, and published?
- What is the typical content lifecycle and how frequently do you review the site for outdated content for updating or removal?
- Is there a governance model in place for content?

Feedback and Data

- What data do you capture about your site and how do you use it? (For example, traffic data, search logs, customer feedback)
- How do you judge website success?

Tagging

- Do you add tags (metadata) to information assets?
- Is metadata tagging a required step in your information-processing workflow?
- Do you use any standard metadata schemes of content attributes?
- Who tags metadata and at what stage in the information lifecycle?
- Do you use any automated or semiautomated tools to tag to information assets?
- Which information assets do you want tagged?
- Which content attributes (title, creator, date, subject, summary, etc.) do you want tagged?
- Do you have legacy content that needs to be tagged?

Categorization

- What website content is easy to find?
- What is difficult to find?
- What are your broad categories of website content? What are the key ways that you need to slice and dice website content?
- Do you use controlled vocabulary lists or classification schemes?
- Do you use internally-controlled naming systems to describe your products and services?
- Who is responsible for developing and maintaining vocabulary lists and naming systems?
- Do you have list of synonyms, abbreviations, and acronyms? Do you have a glossary?

Questions for Site Users

- How do you use the site?
- What is your opinion of the content and features that are relevant to you?
- What's unnecessary?
- What's missing?

Content Audit Checklist

- **Ownership:**
  - ☐ Who owns the content?
  - ☐ Who makes decisions about it?
- **Currency:**
  - ☐ Is content up to date?
  - ☐ Does it reflect current branding, messaging, legal and regulatory requirements, and product and service features?
- **Accuracy:**
  - ☐ Is all the information correct?
- **Relevance:**
  - ☐ Is the information relevant to the audience?
  - ☐ Does it make sense in the context of the content it's related to?
  - ☐ When found out of context (for example, via search or a link), does it make sense?
  - ☐ If not, are there clear pathways to related content or indications of where it lives in the site architecture (for example, breadcrumbs or other navigational clues)?
- **Uniqueness:**
  - ☐ Do other pieces of content cover the same topic?
  - ☐ Of the duplicates, which is more comprehensive, current, and compelling?
  - ☐ Do metrics show which is more visited by users?
- **Brand:**
  - ☐ Does content support your brand guidelines?
  - ☐ Does it tell the organization's story effectively?
  - ☐ Is the tone and voice of the content on brand and appropriate for the intended audience?
- **Call to action:**
  - ☐ Does the content have a clear call to action?
  - ☐ Does the reader know the next step to take?
- **Purpose:**
  - ☐ Does the content fulfill a specific purpose?
  - ☐ What is the business value?
- **Format:**
  - ☐ Is the form of the content effective and appropriate?
  - ☐ Would it be better presented in a different format, such as video, infographic, interactive feature?
  - ☐ Would adding graphics or other media enhance the content?

- **Performance:**
  - ☐ Are there any data about the content (for example, page views, exit and bounce rates, shares) to assess user engagement?
- **Readability:**
  - ☐ Are pages concise and easy to scan?
  - ☐ Is the content written in a style appropriate to the audience and topic?
  - ☐ Is the content free of jargon, acronyms, and other abbreviations – or are those items defined in context?
- **Discoverability:**
  - ☐ Do headings and introductory text include keywords meaningful to users?
  - ☐ Is the content placed appropriately in the site architecture?
  - ☐ Is the page linked from other relevant pages? External pages?
  - ☐ Is the page returned in search results if the user enters keywords related to the main subject of the page?

APPENDIX D
Sample Persona

An example of what a persona for a nonprofit site might look like.

Name: Michael

Role: Trustee, Donor

Occupation: Businessman

Age: 45

Location: Seattle, WA

Wants

- To support organizations that help children and animals in a careful, caring way.

Motivations

- Cares deeply about animal rights and animal lives, is especially motivated by animal rescue and rehabilitation programs

Behaviors

- Skilled online researcher
- Smart phone user
- Attends many fundraisers, sits on the board of many organizations, enjoys meeting with friends who are also activists and donors

Information Needs

- Finding a select group of organizations to support
- Finding moving stories backed by sustainable plans and strong success records
- Evidence of how donated funds are spent
- Painless path to donation
- Features to allow sharing content with friends and contact
- Information about events

APPENDIX E
Customer Journey Map

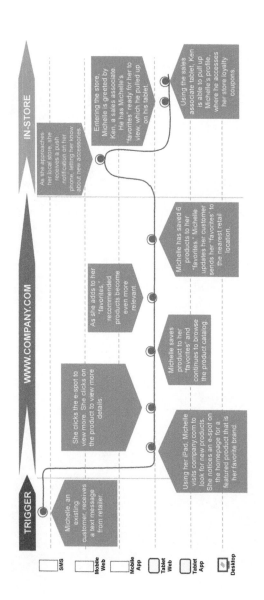

APPENDIX F
Sample Gap Analysis Map

Strategic Content Sample Gap Map

Buyer's Journey

Audience	Awareness	Consideration	Conversion	Loyalty/Advocacy
IT Professional	• Help me recognize I have an issue • Recognize your brand as a leader • Educate me on issues and trends	• Understand value of your product/solution • Get detailed technical information on product/solution • Understand implementation and deployment • Compare to competitive set • Demo or try product or solution	• Help me build a business case • Get third-party validation • Get peer validation (customer stories) • Calculate ROI • Build an RFP or proposal	• Support for product or solution • Training and education • Performance metrics • Review or recommend
Decision Maker/C-Suite	• Educate me on trends/top reports	• Summary of product or solution and business value	• Review business case • Validate decision with peers/analysts • Review ROI/RFP proposal response	• Review performance metrics
Influencer	• Educate me on issues and trends • Insight I can own/blog/publish	• Summary of product or solution and how it meets needs	• Validate my recommendation	• Insider access • Influencer outreach

User Journey

Audience	Awareness	Consideration	Apply	Recommend
Career Seeker	• Recognize your company/brand • Find out if jobs are available for my skills	• Learn why your company is a good place to work • Learn about culture, benefits, work environment • Read peer reviews of company as employer	• Easy, clear way to apply • Get recommendation from peers at company • Research company to prepare for interview • Validate choice	• Ability to share job postings • Share company news

Strategic Content

APPENDIX G
Content Audit Template

The following is an example of how a content audit document might be structured. Like every content set, every audit is unique, so adjust as needed.

Table of Contents

Executive Summary
High-level overview of the project context, goals, and findings.

What We Assessed
Detailed list of content assets that were reviewed, and their sources. Delineate the audit's scope, including what was not assessed.

Business Goals
List and explain the business goals that the content was audited against. Include information gathered in stakeholder interviews, including pain points and wishlists considered as content was reviewed.

User Needs
Describe the audiences and their needs. Introduce personas and customer-journey maps to be referenced in the audit results.

Audit Criteria
Editorial and brand standards, key performance indicators, etc. – the measurements you evaluated content against–and why they matter.

Current State Assessment
The meat of the audit. How did the content perform against the business goals, user needs, and audit criteria? Depending on the audit, this section may be presented section by section, content type by content type, or audience by audience.

Qualitative Audit
Overview of the content quality, with examples. You need not list every example of every issue found; choose representative examples and give a sense of the issue's prevalence and implications.

Content Effectiveness/Performance

Analysis of the content metrics and what they indicate about content effectiveness and user engagement.

Competitive Audit

Comparison of content against top competitors or peers, using a scorecard of criteria against which each was assessed.

Recommendations

Detailed recommendations for content improvement, including recommendations for resources, tools, and processes for carrying out the recommendations and for content governance.

Appendices

Attach supporting documentation, such as your inventory, analytics reports, social-media statistics, etc.

Additional Reading

A selective list of articles and books related to the topics of content inventory and audit. Note that this is a fast-growing discipline and new content is constantly being published, so this list is inevitably incomplete. The authors of these articles and books represent established names in the field as well as newer voices; I encourage you to seek out their work to find even more helpful information.

Content Inventory

[1] Bailie, Rahel. "The Content Inventory."
http://intentionaldesign.ca/2011/02/02/the-content-inventory/.

[2] Bailie, Rahel. "Content Inventories, Audits, and Analyses: All Part of Benchmarking."
http://intentionaldesign.ca/2012/08/09/content-inventories-audits-and-analyses-all-part-of-benchmarking/.

[3] Baldwin, Scott. "Doing a Content Audit or Inventory."
http://nform.com/blog/2010/01/doing-a-content-audit-or-inventory/.

[4] Davis, Sue. "Building the Mother of All Content Inventories."
http://blogs.ec.europa.eu/waltzing_matilda/building-the-mother-of-all-content-inventories/.

[5] Fraser, Janice Crotty. "Taking a Content Inventory."
http://www.drdobbs.com/taking-a-content-inventory/184413339.

[6] Halvorson, Kristina. "The Content Inventory Is Your Friend."
http://blog.braintraffic.com/2009/03/the-content-inventory-is-your-friend/.

[7] Hobbs, David. "Rethinking the Content Inventory: Exploration."
http://hobbsontech.com/content/rethinking-content-inventory-exploration.

[8] Hobbs, David. "Rethinking the Content Inventory: Site Inventories."
http://hobbsontech.com/content/rethinking-content-inventory-site-inventories.

[9] Land, Paula. "The Long Happy Life of a Content Inventory."
http://paula-land.squarespace.com/blog/2012/11/19/the-long-happy-life-of-a-content-inventory.

[10] Martinez, Vanessa. "Everything Old Is New Again: How to Turn Your Content Inventory into an Idea Cache."
http://marketeer.kapost.com/2012/01/everything-old-is-new-again-how-to-turn-your-content-inventory-into-an-idea-cache/.

[11] Maurer, Donna. "Taking a Content Inventory."
http://maadmob.net/donna/blog/2006/taking-a-content-inventory.

[12] Rosenfeld, Lou. "The Rolling Content Inventory."
http://www.louisrosenfeld.com/home/bloug_archive/000448.html.

[13] Veen, Jeffrey. "Doing a Content Inventory (Or, a Mind-Numbingly Detailed
Odyssey through Your Web Site)."
http://www.adaptivepath.com/ideas/doing-content-inventory.

Content Audit

[14] 4Syllables. "Content Audit Guide and Template."
http://www.4syllables.com.au/resources/templates-checklists/content-
audits/.

[15] Detzi, Christopher. "From Content Audit to Design Insight: How a Content
Audit Facilitates Decision-Making and Influences Design Strategy."
http://uxmag.com/articles/from-content-audit-to-design-insight.

[16] Evans, Susan. "Content Inventory + Content Audit: Why Would You Do
One without the Other?"
http://susantevans.wordpress.com/2012/07/07/content-inventory-
content-audit-why-would-you-do-one-without-the-other/.

[17] Fasso, Tosca. "Audits to Insights"
http://content-insight.com/blog/2013/08/audits-insights-part-1/.

[18] Land, Paula. "Website Content Audits."
http://content-insight.com/resources/content-inventory-and-audit-
articles/website-content-audits/.

[19] Leibtag, Ahava. "Why Traditional Content Audits Aren't Enough."
http://contentmarketinginstitute.com/2011/01/content-audits/.

[20] Marsh, Hilary. "How to Do a Content Audit."
http://www.hilarymarsh.com/.

[21] McCarthy, Rob. "How To: Conducting a Website Content Audit."
http://www.cmswire.com/cms/customer-experience/how-to-
conducting-a-website-content-audit-013097.php.

[22] Spencer, Donna. "How to Conduct a Content Audit."
http://uxmastery.com/how-to-conduct-a-content-audit/.

[23] Wachter-Boettcher, Sara. "Content Knowledge Is Power."
http://www.smashingmagazine.com/2013/04/29/content-knowledge-
is-power/

Analytics

[24] Colman, Jonathon. "Data Sets You Free: Analytics for Content Strategy."
http://www.slideshare.net/jcolman/data-sets-you-free-confab-2013.

[25] Nichols, Kevin P. and Rebecca Schneider. "Positioning Content for Success:
A Metrics-Driven Strategy."
http://www.sapient.com/en-us/sapientnitro/thinking/paper/128/
positioning_content_for_success_a_metricsdriven_strategy.html.

Content Governance

[26] Allen, Rick. "Planning for Content Governance."
http://www.slideshare.net/epublishmedia/planning-for-content-governance.

[27] Welchman, Lisa. "Web Governance: A Definition."
http://welchmanpierpoint.com/blog/web-governance-definition.

[28] Earley, Seth. "Developing a Content Maintenance and Governance Strategy."
http://www.asis.org/Bulletin/Dec-10/DecJan11_Earley.html.

Personas and Customer Journey Maps

[29] Mausser, Kristina. "Why Personas Are Critical for Content Strategy."
http://johnnyholland.org/2012/02/why-personas-are-critical-for-content-strategy/.

[30] Nichols, Kevin P.. "Personalization, Customer Journey, Omnichannel: A
How-To Approach."
http://www.slideshare.net/contentstrategyworkshops/cs-workshopomnichannelpersonalizationcontentstrategyknichols.

[31] Risdon, Chris. "The Anatomy of an Experience Map."
http://www.adaptivepath.com/ideas/the-anatomy-of-an-experience-map/.

Legal Issues Resources

[32] BitLaw. *Website Legal Issues.*
http://www.bitlaw.com/internet/webpage.html.

[33] Executive Counsel. *Benefits of a Legal Audit.*
http://www.exec-counsel.com/content/benefits-legal-audit.

[34] Nonprofit Resource Center. *Nonprofit Resource Center.*
http://www.nprcenter.org/legal-regulatory.

[35] Nonprofit Law Blog. *Top 5 Fundraising Legal Tips.*
http://www.nonprofitlawblog.com/home/2010/09/top-5-fundraising-legal-tips.html.

Books

Most books related to content strategy and content management include sections on the value and process of doing content inventories and audits. A few of the best known are listed here along with a few specialized topics.

[36] Abel, Scott and Rahel Bailie. *The Language of Content Strategy*. XML Press. 2014.

[37] Bailie, Rahel and Noz Urbina. *Content Strategy: Connecting the dots between business, brand, and benefits*. XML Press. 2013.

[38] Halvorson, Kristina. *Content Strategy for the Web*. New Riders. 2010.

[39] Rockley, Ann and Charles Cooper. *Managing Enterprise Content: A Unified Content Strategy*. 2nd ed. New Riders Press. 2012.

[40] Singh, Nitish and Arun Pereira. *The Culturally Customized Web Site: Customizing Web Sites for the Global Marketplace*. Routledge. 2011.

[41] Yunker, John. *Beyond Borders: Web Globalization Strategies*. New Riders. 2002.

Glossary

analytics

Website traffic and usage statistics, typically capturing data such as how many users have viewed a page, how long users stay on pages, the pages through which users enter or exit a site, and the paths through which users traverse the site.

competitive audit

A content audit conducted for the purpose of comparing similar sites, done by selecting a set of common content types or functionality and ranking sites against one another.

content audit

The process and result of conducting a quantitative study of a content inventory.

content inventory

The process and result of creating an organized listing of content assets (text, files, audio, video, images) for a body of content. An inventory includes as much information about each piece of content as possible.

content lifecycle

The process that defines the series of changes in the life of any piece of content, including reproduction, from creation onward.

content management system

A software application that supports information capture, editorial, governance, and publishing processes with tools such as workflow, access control, versioning, search, and collaboration.

content matrix

An expansion of the content inventory to track the progress of each piece of content through the stages of a project or content lifecycle.

content migration

The one-time movement of content from one repository to another.

content strategy

The analysis and planning to develop a repeatable system that governs the management of content throughout the entire content lifecycle.

conversion
A measure of a desired user action, such as making a purchase or registering for an account

customer journey map
A representation of a user's step-by-step content interactions during task completion, both online and offline. The map documents the content and experiences at each point and describes the user's state of mind and information needs at each step.

DAM
Digital asset management. The process and technology used to store and manage digital assets such as images.

gap analysis
The process of comparing current state, usually as assessed in a content inventory and audit project, with a future state, and analyzing the differences and the activities and resources needed to create the future state.

globalization
The analysis of, and planning for, the development, delivery, and consumption of global content; in essence, it is the analysis which forms a global content strategy.

governance
The systems, policies, and processes used to manage and control a content set to ensure consistency, efficiency, and compliance with standards.

heuristics
Measures by which content can be evaluated, generally based on experiential factors.

interaction model
A representation of how an application or content set is used, typically expressed as a flow chart or other type of diagram.

key performance indicator
A metric selected by an organization to evaluate success and track progress toward goals. Examples include sales quotas, customer satisfaction ratings, social media engagement levels.

localization
Adaptation of content to make it more meaningful, appropriate, and effective for a particular culture, locale, or market.

multichannel

> A multichannel content strategy addresses the various publication or distribution points at which content will be accessed by users and ensures that the experience is relevant in each context.

multidevice

> Content that is designed to appear appropriately formatted for different devices. May also mean that content is curated to be relevant to the context in which the device is used.

persona

> A fictional representation of a user type, based on user research and customer data, that is used as a proxy against which content and features can be tested. Personas usually include a photograph, demographic information, and a description of the person's content needs and behaviors.

qualitative audit

> A content audit that focuses on editorial quality of content, including aspects such as consistency, currency, relevance, tone, and voice.

RACI

> A method of assigning roles and responsibilities to the people involved in a project. RACI stands for Responsible, Accountable, Consulted, Informed.

RAITES method

> An audit rubric designed by Rahel Bailie for evaluating if content is Relevant, Accurate, Informative, Timely, Engaging, and Standards-Based[1]

redirect

> A technical function that switches the resulting web address from the one that was clicked or input by the user to another web location. Generally used to address URL changes and avoid broken links.

return on investment

> A calculation of whether the end result of an action or project sufficiently pays off the investment in time and resources to achieve the result.

ROT

> A set of qualities (Redundant, Outdated, Trivial) against which content can be assessed. Coined by Lou Rosenfeld.

---

[1] http://intentionaldesign.ca/2011/09/28/content-that-raites/

site map
> A visual representation of the structure of a site.

taxonomy
> A hierarchical classification scheme made up of categories and subcategories of information plus a controlled vocabulary of terms, usually used to describe a specific area of knowledge.

user flows
> The paths through which a user might traverse the content and functionality of a website or content set. Often represented in flow charts.

.

About the Author

Paula Land is a content strategy consultant and technology entrepreneur. As founder and principal consultant at Strategic Content, she develops content strategies and implementation plans for private clients ranging from nonprofits to large e-commerce sites. As cofounder of Content Insight,[1] she is the impetus behind the development of CAT, the Content Analysis Tool, which creates automated content inventories.

Before founding her own businesses, Paula was a user experience and content strategy lead for Razorfish, a leading digital agency, where she led the content strategy on the development of enterprise-level websites, redesigns, and CMS implementations.

Paula has worked for over twenty-five years in content-related roles, spanning all aspects of the content lifecycle, with a focus on delivering large-scale, complex websites.

Paula is a frequent speaker at conferences and presents workshops and webinars on the topic of content inventories, audits, and analyses. She contributed the essay on content inventory to the book *The Language of Content Strategy*.

---

[1] http://www.content-insight.com/

Index

Colophon

About the Book

This book was authored, edited, and indexed in a Confluence wiki. Contents were exported to DocBook using the Scroll DocBook Exporter from K15t Software. The book was then generated from that output using the DocBook XML stylesheets with XML Press customizations and, for the print edition, the RenderX XEP formatter.

About the Content Wrangler Content Strategy Book Series

The Content Wrangler Content Strategy Book Series from XML Press provides content professionals with a road map for success. Each volume provides practical advice, best practices, and lessons learned from the most knowledgeable content strategists in the world. Visit the companion website for more information contentstrategybooks.com.

About XML Press

XML Press (xmlpress.net) was founded in 2008 to publish content that helps technical communicators be more effective. Our publications support managers, social media practitioners, technical communicators, and content strategists and the engineers who support their efforts.

Our publications are available through most retailers, and discounted pricing is available for volume purchases for educational or promotional use. For more information, send email to orders@xmlpress.net or call us at (970) 231-3624.

The Content Wrangler
Content Strategy Book Series

The Content Wrangler Content Strategy Book Series from XML Press provides content professionals with a road map for success. Each volume provides practical advice, best practices, and lessons learned from the most knowledgeable content strategists in the world.

The Language of Content Strategy

Scott Abel and Rahel Anne Bailie

Available Now

Print: $19.95
eBook: $16.95

The Language of Content Strategy is the gateway to a language that describes the world of content strategy. With fifty-two contributors, all known for their depth of knowledge, this set of terms forms the core of an emerging profession and, as a result, helps shape the profession. The terminology spans a range of competencies with the broad area of content strategy.

Content Audits and Inventories: A Handbook

Paula Ladenburg Land

Available Now

Print: $24.95
eBook: $19.95

Successful content strategy projects start with knowing the quantity, type, and quality of existing assets. Paula Land's new book, *Content Audits and Inventories: A Handbook*, shows you how to begin with an automated inventory, scope and plan an audit, evaluate content against business and user goals, and move forward with a set of useful, actionable insights.

Enterprise Content Strategy: A Project Guide

Kevin P. Nichols

Available Summer, 2014

Print: $24.95
eBook: $19.95

Kevin P. Nichols' *Enterprise Content Strategy: A Project Guide* outlines best practices for conducting and executing content strategy projects. His book is a step-by-step guide to building an enterprise content strategy for your organization.

Global Content Strategy: A Primer

Val Swisher

Available Winter, 2014

Print: $24.95
eBook: $19.95

Nearly every organization needs to serve customers around the world. This book describes how to build a global content strategy that addresses analysis, planning, development, delivery, and consumption of global content that will serve customers wherever they are.

ContentStrategyBooks.com
XMLPress.net

CPSIA information can be obtained at www.ICGtesting.com
Printed in the USA
BVOW11s1016250914

368130BV00009B/58/P